Contents

ACKNOWLEDGEMENTS TO PHOTOGRAPHS

Photographs are acknowledged as follows: Roy Armstrong, plates 18, 19, 20, 21, 24; National Monuments Record, plates 2, 3, 4, 9, 14, 15; Pamla Toler, plate 1. Other photographs are by the author.

 First published in 1978. Second edition 1979; reprinted. No. 242 in the Discovering series. ISBN 0 85263 481 1.

Printed by C. I. Thomas & Sons (Haverfordwest) Ltd., Press Buildings, Merlins Bridge, Haverfordwest, Dyfed.

RICHA ARRIS

Discovering Timber-framed Buildings

SHIRE PUBLICATIONS LTD

Introduction

'Half-timber' and 'black and white' are the common names for timber-framed buildings and describe what is perhaps the most delightful and characteristic feature of many parts of the English and Welsh countryside and historic towns. Nobody knows how the term 'half-timber' arose, whether from the practice of halving trees or from the half timber and half plaster construction of external walls, but 'black and white' refers to the layers of black and white paint applied to timbers and panels (in very many cases a recent 'improvement', unfortunately).

This 'magpie' surface is very familiar, but there is much more to these buildings than their exterior. It is the 'wealth of old beams' – so beloved of estate agents – that we are concerned with here. What exactly are these beams? How do they make a building? Exposed beams still mean home to people in a way that plastered walls and ceilings never will, but their meaning and purpose is forgotten. The fibreglass beams installed in some 'olde worlde' pub interiors, for example, are to many people attractive and welcoming but would be hard put to it to support a hen coop.

The purpose of this book is to show how beams were put together to form buildings. Buildings – at least those which survive today – were not home-made but were produced by carpenters who had served a long apprenticeship to learn the skills of their craft. Creating a building from trees is a bit like alchemy. Instead of turning base metal to gold, the alchemist-carpenter had to turn trees into beams, into frames, into buildings.

The secret of this magic was the craft tradition. This gave the carpenter a series of clear steps by which he could find his way through the maze of difficulties he faced in each new building. Each step gave a key to a part of the process. Any change in these essential steps would upset the balance of his craft, but around them he was able to create the unique character of each building.

In most areas this craft tradition came to an abrupt end in the late eighteenth or early nineteenth century. Since our earliest surviving timber-framed buildings date from the thirteenth century we have the product of six hundred years to consider. Timber buildings of the thirteenth and fourteenth centuries have been given a considerable amount of attention by historians, but they are rare and the reader is unlikely to discover any which are not already known. They have therefore been given little space in this book. On the other hand, surprisingly little has been published to describe the buildings which are no less fascinating but much more numerous and easy to find: houses, cottages and barns built in the sixteenth, seventeenth and eighteenth centuries. This book is intended as an introduction to the subject. The technical terms will be unfamiliar to many people, but reference to the drawings and the glossary-index should resolve most difficulties.

The author would like to thank all those people who have helped in the preparation of this book, and in particular the following: Roy Armstrong; F. W. B. Charles; David Michelmore; and members of the Vernacular Architecture Group. Most of the drawings are based on measured surveys made by the author, whose thanks are due to the many owners of buildings in which he has clambered about, tape measure in one hand, pencil and paper in the other.

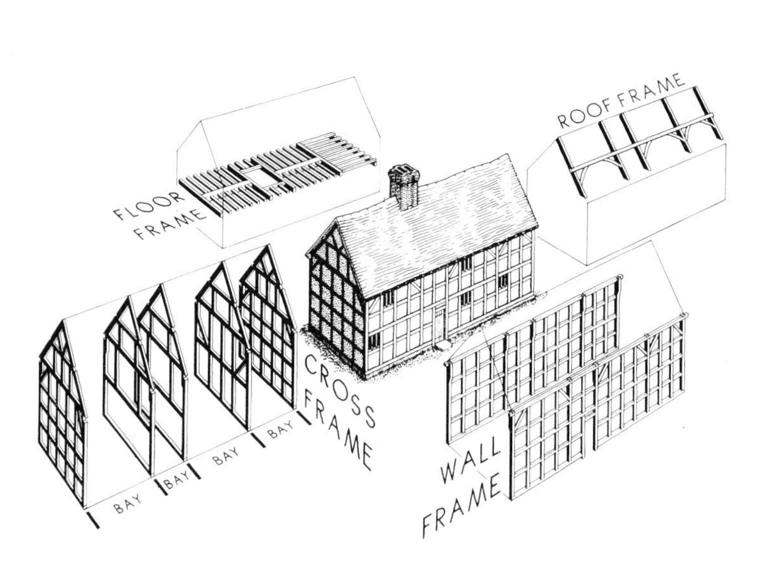

Fig. 1. Timber-framed buildings: bays and frames.

1. Bays, frames and boxes

The idea of using building blocks is familiar to everyone from childhood and it is easy to assume that piling one block on top of another is the only way of building. Cosy brick cottages, cathedrals and Greek temples all use the same method, relying on gravity to maintain brick on brick or stone lintel on stone column. Timber-framed buildings, however, use an entirely different method: the framing members (posts, beams, studs, rails, plates and rafters) are held together by being *jointed* to one another. Joints are independent of gravity: a giant could pick up a timber-framed house, turn it upside down and put it back on the ground still in one piece.

Timber-framed buildings are intricate structures and were always prefabricated: all joints had to be cut and fitted before any building could begin. In prefabrication the carpenter treated the building as a series of **frames**. The side-wall frames, cross-wall frames and roof trusses, as well as floor frames (of beams and joists) and roof frames (of purlins and wind braces), had to be carefully fitted together (framed up) flat on the ground before the erection of the building. When erected the frames were joined together as the sides and subdivisions of a box, forming the skeleton which was then covered or infilled to form the walls, roof and floor.

Bays

It is often difficult to chart the history of a building from original design through subsequent alterations to the present day, but the original **bay** divisions can give a simple key to understanding the plan and structure. The idea of a bay of a building is still familiar to us: modern concrete and steel buildings are built in bays, as were cathedrals and churches. Part of the idea is to collect the structural loads at certain points where main columns are placed. In timber-framed buildings the bay divisions are also points at which the building is tied together across its span.

The relationship between the plan of a house and the bay system of the structure was usually simple enough, each bay forming one room. Sometimes, however, a cross frame is 'open' on one floor and 'closed' on the other so that, for example, a two-bay chamber may lie above two single-bay ground-floor rooms. In many cases intermediate roof trusses are introduced to halve the structural length of a bay. Medieval open halls are often of two bays with a central open arched cross frame or roof truss. In houses which have been much altered it is often necessary to clamber (carefully) into the roof space to discover the original bay divisions from the roof trusses.

Medieval timber-framed buildings were often described as being of one, two, three or four bays, in legal documents for instance, but this did not mean that all bays were the same length. It gave a convenient measure of size in much the same way as modern houses are described by the number of bedrooms they contain. The lengths of bays vary between about 5 feet and about 20 feet (1.5 to 6 metres). Very short bays were usually designed to locate a cross-passage, a smoke bay, a smoke hood or chimney stack. No general rule applies to the rest of the range of bay sizes but a local researcher will soon get to know the lengths to expect in buildings of a particular period.

It is sometimes supposed that bay lengths were determined by the length of timber available, but this would only have defined a maximum and probably had little influence in practice. In his book *The Evolution of the English House* S.O. Addy suggested (and his suggestion has often been repeated) that the frequent occurrence of bays about 16 feet (4.9 metres) long could be related to the space necessary to stall two pairs of oxen. The reader will be able to test this theory for himself in measuring byres and barns.

Cross frames and roof trusses

Much of the description of timber-framed buildings, in this book and elsewhere, is conducted in terms of cross frames and, in particular, roof trusses (which form part of cross frames). The design of these frames, which changed gradually over the years, helps us to define the character and period of construction of the building. This is partly because it was in roof construction that the most easily recognisable technical changes took place and roof trusses are, therefore, one of the most important features from which a house can be dated. The cross frames which divided room from room in traditional houses were not like the plain anonymous partitions we build nowadays: they were, in a sense, 'live' rather than 'neutral' walls, and their design had a direct bearing on the character and interrelationship of the space they defined.

Each of the following sections describes one of the main types of cross frame: **post and truss**, **aisled**, and **cruck**. It is necessary to introduce here a number of technical terms which will be used in the rest of the book. If they are unfamiliar to you, do not try to learn them all at once, but remember to refer back to this page when you need to. Some of the terms are still used by builders, and others have been generally adopted to describe types of structure which have no modern equivalent. It is useful to remember that most forms of roof construction are named after the most distinctive member of the roof truss: e.g. crown-post roof.

Post and truss cross frames

By far the most common type of cross frame may be termed 'post and truss'. The **main posts** in the outside walls support the **wall plates** (the horizontal beam forming the top of the walls) and are joined at eaves level by a **tie beam**. This forms the base for a **roof truss** which supports the roof **purlins**. Purlins are longitudinal members in the roof giving support to the **common rafters**. Roof trusses very often consist of a pair of **principal rafters** tenoned into the ends of the tie beam and joined by a **collar**.

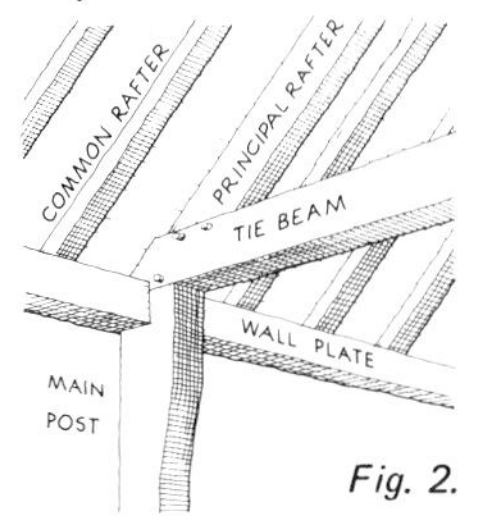

Fig. 2.

The design of roof trusses varies in different places and at different periods and the most common forms will be described in this book under the regions in which they occur. The following is a summary of the basic forms.

Trusses with heavy principal rafters: the principal rafters are much larger than the common rafters; typically found in the same areas as crucks (see below), they may be partly derived from them. The purlins are normally **trenched** into the backs of the principal rafters, but can also be tenoned or threaded.

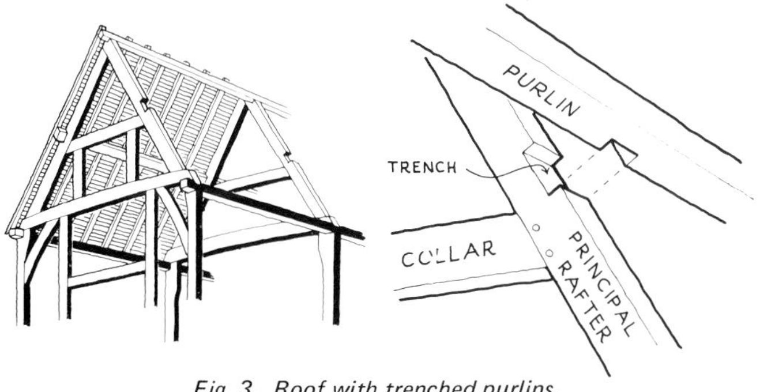

Fig. 3. Roof with trenched purlins.

Trusses with light principal rafters: the principal rafters are similar to common rafters in size, and the purlins are usually **clasped** between principal rafters and a collar or a pair of struts. Found in most parts of the country but probably originated in lowland England (the east and south-east).

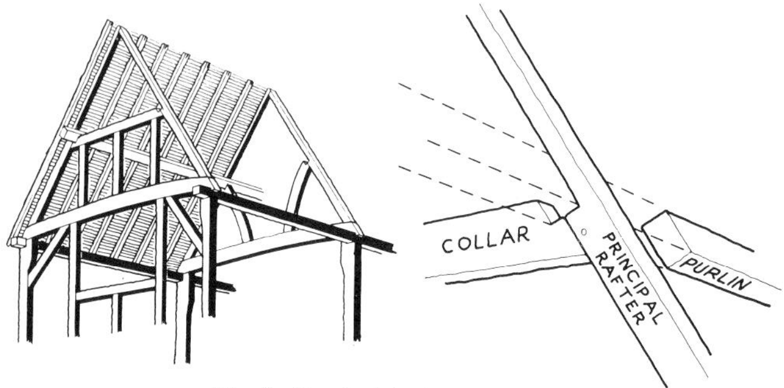

Fig. 4. Roof with clasped purlins.

Trusses with crown posts: instead of side purlins some roofs have a central beam known as a **crown plate** (because it sits on the **crown posts**). Each pair of common rafters is joined by a collar which rests on the crown plate. This type of roof is characteristic of medieval buildings in lowland England. An alternative name sometimes used for the crown plate is *collar purlin* (because it is a purlin which supports collars).

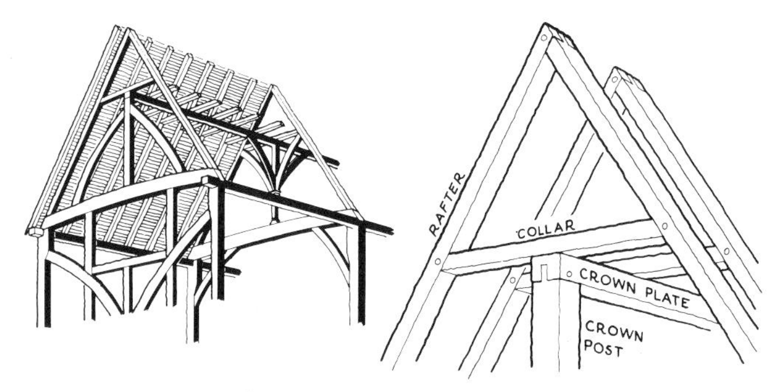

Fig. 5. Roof with crown plate.

Crucks

Crucks are long curved timbers, framed together in pairs and joined by a tie beam or collar, which rise from ground level to support the roof purlins of a building. They are very numerous, with well over two thousand examples recorded in England and Wales. Their geographical distribution is very surprising, as they are completely unknown in the eastern and south-eastern lowland of England but are fairly evenly distributed over most of the rest of the north, south, west and midlands of England and Wales.

Crucks have often been supposed to have been developed gradually from a primitive form of construction in which two poles were leaned against each other and thatched, but there is no direct evidence to support this theory. Most cruck buildings of early date are extremely well developed and sophisticated in their carpentry and it is quite possible that British cruck construction was *invented* at a high level in society, perhaps during the twelfth century, and later descended the social scale to be used in smaller halls, cottages and barns. Much energy has been expended in forming various theories of the origin of crucks but so little evidence survives of buildings earlier than about 1200 that the true answer will never be known.

Fig. 6. Crucks: interior of a Herefordshire barn.

Aisled buildings

Roof trusses can only span across a limited width. If larger areas of ground plan are needed they can be achieved by dividing the structure into a central **nave** with side **aisles.** The nave is roofed with trusses of one of the kinds already described, supported on posts which are now internal **arcade posts** (rather than external wall posts), and the aisles are roofed with subsidiary lean-to roofs. There are about ninety examples of aisled timber-framed halls known in England and Wales, mostly of medieval origin and surviving in a much altered condition. However, aisled construction was used over a much longer period for barns and many examples can be seen, particularly in East Anglia, southern England, and parts of Yorkshire.

The internal posts in an aisled building obviously obstruct the space in halls and barns, and to overcome this problem medieval carpenters invented special types of structure. The problem they faced was the support of the **arcade plate**, the longitudinal plate – or beam – which normally sits on the arcade posts (i.e. the equivalent of the wall plate in post and truss buildings without aisles). One solution was to support it on a **base cruck** – essentially a large curved post inclined inwards. Another was to support it on a short post (the

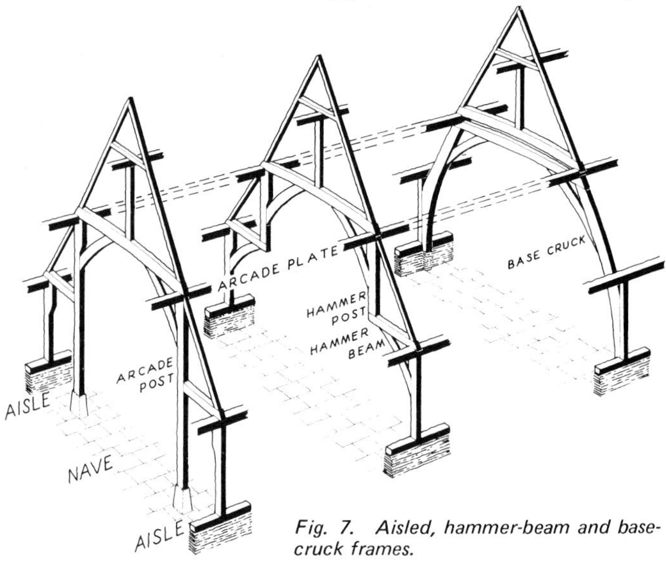

Fig. 7. Aisled, hammer-beam and base-cruck frames.

hammer post) which rested on a short beam (the **hammer beam)** cantilevered in from the top of the outside wall and braced against it. The enormous roof of Westminster Hall in London is based on the hammer-beam principle. Base-cruck and hammer-beam roofs were only used in buildings of high status (great halls and barns). Both had a rather distinctive appearance, and hammer beams continued to be used in modified form in great halls long after aisled construction had gone out of fashion.

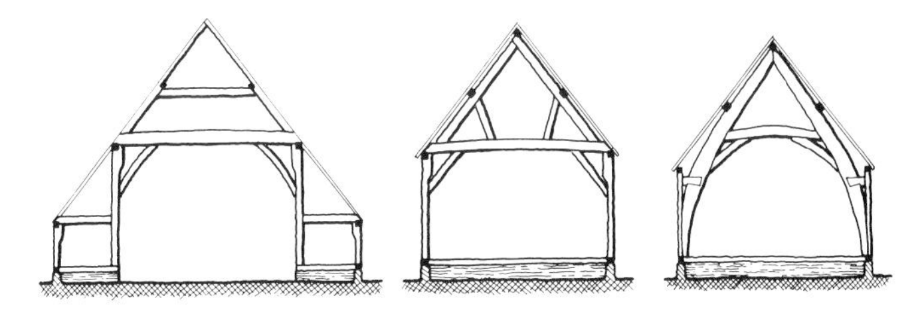

Fig. 8. Cross frames: aisled, post and truss, cruck.

Highland and lowland zones

In studying traditions of timber-framed building in England it is helpful to keep in mind a picture of two main zones or regions. The so-called *lowland* zone may be thought of as the area to the south and east of the belt of Jurassic limestone which runs from Portland Bill, via the Cotswolds, to Lincolnshire, but including also the eastern plains of Lincolnshire and east Yorkshire. The *highland* zone comprises the remainder: the south-west, the midlands and the north. The distribution of crucks, in the highland zone, and of aisled buildings and crown-post roofs, in the lowland zone, most closely corresponds with this division, albeit with areas of overlap. It is also useful in considering the dominance of different patterns of wall framing and of house plans. In agriculture the two zones correspond, although very roughly, with areas of pastoral-dominated farming (highland) and arable-dominated farming (lowland), the result of differences of climate and terrain. Within these zones the institutions and organisation of rural society – the manor, the village, the farm and the family – varied widely. What is, perhaps, surprising is the way in which building traditions seem to have been tied more closely to the underlying farming conditions than to the superstructure of social organisation.

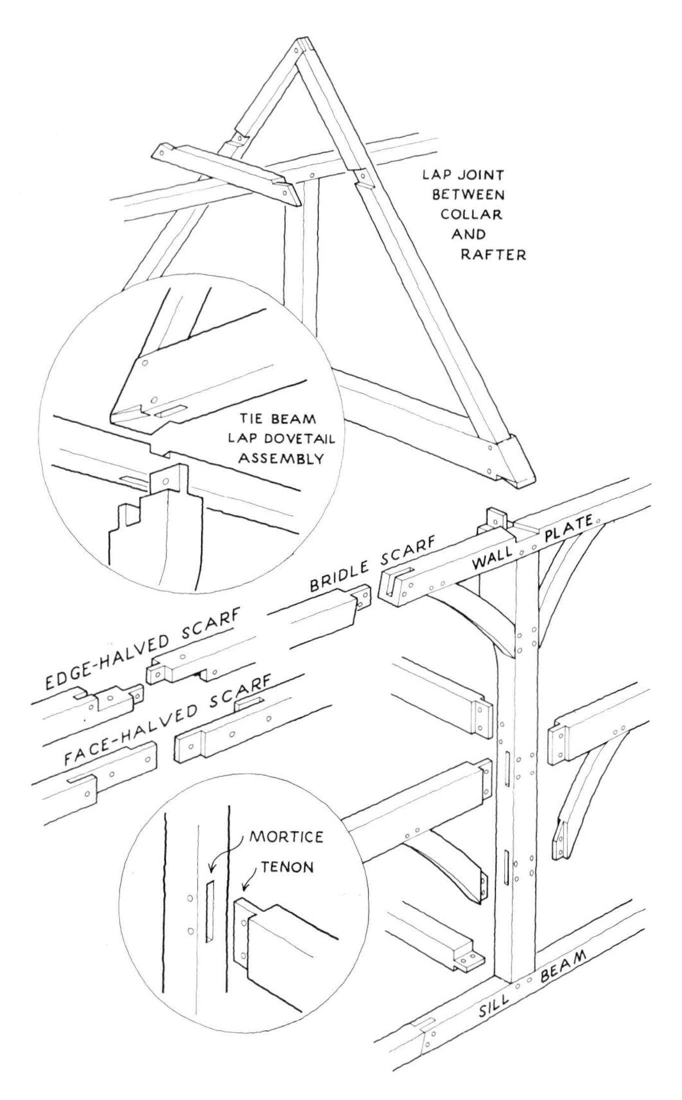

Fig. 9. Joints in traditional carpentry.

2. Construction

Joints

Modern timber buildings are assembled with nails and metal timber connectors: they contain few, if any, joints. By contrast, in traditional timber-framed buildings members come together in timber joints: mortice and tenon, and half-lap. Similar joints can be seen in any piece of traditional furniture or joinery work. A third kind of joint, the scarf, is used in buildings to join two beams together in line to form a continuous member – mainly purlins, wall plates and sill beams – because trees are seldom as long as buildings.

Mortice and tenon joints were the basis of all traditional framing and even a quite small building will contain a very large number of them: collectively they stiffen the whole frame. The existence of mortice and tenon joints is a guide to whether a particular member was part of the original building, since a beam which is tenoned at both ends cannot be inserted into an existing framework. Mortice and tenon joints are almost always secured with oak pegs, and when a beam has been removed the empty peg holes give a useful clue to its former existence.

Mortice and tenon joints vary hardly at all in design but many different forms of **scarf joint** can be found: three basic types are shown in fig. 9. Scarf joints often occur near a main post, and it is usually possible to work out from their positions the order in which the bays of the building were erected.

Lap joints were used in relatively few positions in timber-framed buildings. They were a significant element in early medieval forms of construction, but in later buildings they are generally found only between rafters and collars (in crown-post roofs) and in cruck construction. The most important lap joint, however, was the **tie beam lap-dovetail** assembly (see fig. 9). This complicated but beautiful detail was extraordinarily long-lived (from the thirteenth century until the nineteenth), in almost universal use in all buildings (except crucks), and is fundamental to the British tradition of timber framing. In *The City and Country Purchaser* of 1726, a builder's price book and dictionary, Richard Neve defined the word 'beam' in terms of the lap-dovetail assembly – the joint was evidently an indispensable part of building technique.

> *Beam*: In Building is a piece of Timber, which always lies cross the Building, into which the Feet of the principal Rafters are framed; no Building hath less than 2 of these Beams, viz. one at each Head; into these Beams the Girders of the Garret-floor are framed; and if it be a Timber-Build-

ing, the Teazle Tennons of the Posts are framed. The Teazle Tennons are made at right angles to those which are made on the Posts to go into the Raisons, and the Relish, or Cheats of these Teazle Tennons stand up within an Inch and a half of the top of the Raison; and the Beam is cauked down (which is the same as Dove-tailing a-cross) till the cheeks of the Mortices in the Beam conjoin with those of the Teazle Tennon on the Posts.

All of the three basic kinds of joints were already in use in the earliest buildings which still survive, and the mortice and tenon and the half-lap remained essentially the same for several centuries. Sometimes a new joint or assembly was invented to solve a specific problem (for instance the framing of upper floors or of jetties) and there were many changes made in detailed design, particularly of scarf joints, which are interesting to record. However, it is generally misleading to analyse traditional carpentry using modern structural analysis. Joints were only one aspect of the slowly changing traditional package of plan, form and construction, which depended for its success more on the compatibility of its constituent parts than on the economy or efficiency of each element.

In marking out a joint for a frame of a building the carpenter always had to choose a face of the timber from which to work: during prefabrication, when the frame was fitted together on the ground, this face would be uppermost and is therefore known as the **upper face** (fig. 10). In external walls the upper face is the external face; in internal frames its position varies but always bears a traditional relationship with the plan of the building as a whole. For instance, in three-bay barns the upper faces of the two internal frames face inwards towards the threshing floor. After final assembly wooden pegs were driven through holes pre-drilled through each joint: the head

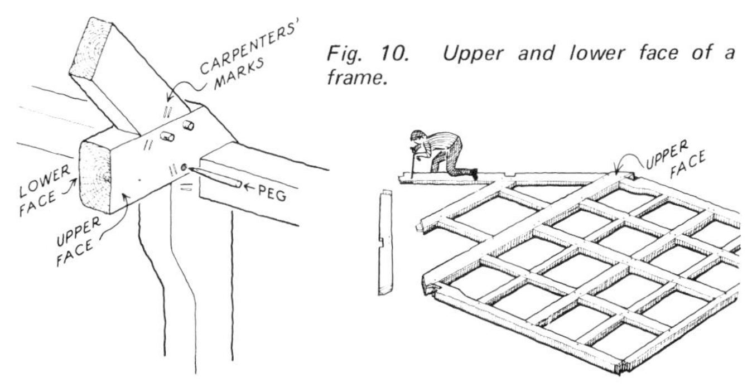

Fig. 10. Upper and lower face of a frame.

of the peg will be seen on the upper face, the pointed end on the lower face, because they were always driven through from the upper face.

Prefabrication of a building often took place in a carpenter's yard or 'framing ground' some distance from the building site: Westminster Hall roof, for instance, was prefabricated at Farnham in Surrey and the enormous timbers were taken by cart and barge to Westminster. Because of this the various pieces had to be identified in some way, and modified Roman numerals were nearly always used, scribed or chiselled on the upper face of the timber. Two basic systems of numbering are found, one numbering members, the other numbering frames. Of these the more common is the former, in which each mark shows the number of a member within a frame, together with a tag to identify the frame to which it belongs (fig. 11). The alternative system was used only in cross frames: each frame is allotted a number which is scribed on each of the main members within it, tags being used to distinguish left from right. For example, in cross frame *III* one principal rafter will be marked *III*, the other *IIʎ*. Carpenters' marks can give a

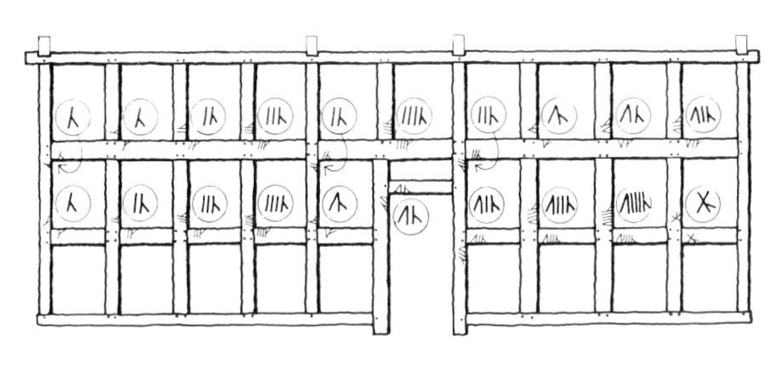

I II III IIII Λ ΛI ΛII ΛIII ΛIIII X XI XII XIII XΛ XΛI XX XXI XXΛ XXX
1 2 3 4 5 6 7 8 9 10 11 12 13 15 16 20 21 25 30

1 2 3 1 2 3 1 2 3 1 2 3 1 2 3 1 2 3 1 2 3

Fig. 11. Carpenters' marks. Top: outside of the wall frame of an eighteenth-century Herefordshire cottage. The numbers each carry a single tag to identify the frame. The short upper studs are marked 1–7, the girding beams (upper rails) 1–3 at their left-hand ends, and the lower rails 1–10 also at their left-hand ends. The wall plate and sill beams are not marked. Middle: a typical sequence of marks showing modification of Roman numerals. Bottom: a variety of identification tags attached to numbers 1, 2, 3.

good deal of information about a building (whether part of it is missing, for instance) and are well worth observing and recording.

Assembly of the frame

After prefabrication of the frames, with all joints cut and snugly fitted, the building had to be erected on site ready to be finished and occupied. Sometimes whole frames would be reared, already assembled, into position: certainly many cruck frames were erected in this way, and perhaps also the wall frames in post and truss buildings. With a building of any size many strong arms would be needed for the rearing and building accounts sometimes included special mention of payments and refreshments distributed on such occasions. However, the majority of members in all buildings must have been assembled one by one, the assembly proceeding in a definite order which can often be guessed by careful consideration of the joints (especially scarf joints) in a frame.

Most of the assembly work can be done using only a rope and pulley; for the larger jobs, such as rearing a frame, shearlegs may have been used as a form of crane. Almost all beams can be carried by one or two men, and each floor level can be used as a platform to facilitate assembly of the higher levels. Scaffolding would also have been used, consisting of poles lashed together, with woven wattle hurdles (instead of planks) as platforms.

Early timber buildings had their main posts dug into the ground, but almost all buildings which still survive were built up above the ground, the lowest member of the frame being a **sill beam** into which the main posts and other uprights were framed. During erection of the frame this beam may have been supported on temporary blocks, replaced after completion

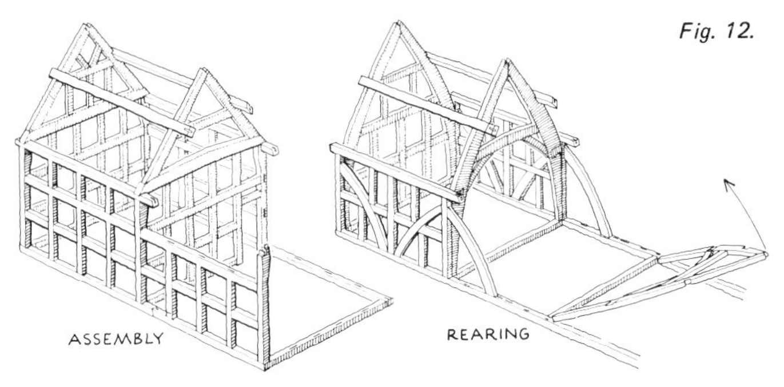

Fig. 12.

with a low stone or brick plinth wall to keep the beam away from the damaging damp ground, a method which ensured an exact fit between the frame and the plinth wall.

Timber and tools

The problem confronting the traditional carpenter was that of transforming trees into beams without the benefit of mechanical sawmills (at least until the mid eighteenth century). The most common methods of conversion produced from each log either one beam of roughly square section (boxed heart), the edges being squared by axe or adze, or two beams of rectangular section, the log being sawn or split in half. For planks and thin studs logs were usually sawn through and through (slabbed), and sometimes the four quarters of a straight log were used for four main posts. As to length, beams of 10-20 feet (3-6 metres) are usual; lengths of over 30 feet (9 metres) are not uncommon, and medieval carpenters could occasionally find beams up to 50 feet (15 metres) long.

Fig. 13. Conversion of timber.

Sawing of beams was done either over a pit or on trestles, the saw being controlled by two men, one above and one below. This is a slow and laborious task but many timbers still bear the resulting irregular saw marks as witness to the method used. Cleaving logs with wedges is very much quicker than sawing: the results with large logs can be rather unpredictable, but poles, laths, staves and pegs would all have been cleft and split rather than sawn.

Although scantlings were used which were heavier than we would now consider necessary, the techniques of conversion and the permitted standards of finish allowed use to be made of trees far smaller than can produce usable timber in modern sawmills. Thus a small oak tree 8-10 inches (200-250 millimetres) in diameter can be split in half and cleaned down to give two excellent rafters approximately 5 inches by 3 inches (125 x 75 millimetres) in section: because the split follows the grain of the tree there is no danger of short grain or excessive amounts of sapwood as there would be in sawing the same log. For larger scantlings it was the usual practice to squeeze the beam out of the smallest possible tree, so that sapwood and even bark often remain on the 'lower' face and edge of the timber: all that was needed in the case of a main post, for instance, was two clean faces of the correct dimensions.

Trees do not always grow straight, and the traditional carpenter must have constantly faced the problem of making maximum use of irregular timber. A number of possibilities were open to him. Small trees with slight irregularities could be split in half to make rafters, as described above, the split being arranged so as to be as straight as possible rather than attempting to go round the curves (fig. 14). Similarly, larger trees could be split or, more likely, sawn in half to make the heavy floor joists which are found in medieval buildings – typically 8 by 6 inches (200 by 150 millimetres) in section. Again the saw cut would be placed so as to produce two mirrored halves of the tree: cutting through the curve would produce timbers thick in some places and thin in others. Trees with a considerable curve could be cut in half for use as tie beams, where the resulting upward camber produces a satisfying aesthetic effect; for smaller members such as braces the two matching mirror-image halves of the log are often found placed symmetrically in a frame. In cruck construction also each pair of crucks very often consists of the two halves of a bent tree, the crucks matching each other and producing a symmetrical frame.

To modern builders it seems odd that joists and rafters in traditional buildings, particularly in the fourteenth and fifteenth centuries, were usually placed 'flat' – that is, with their smaller dimension vertical – rather than 'on edge'. Can medieval carpenters have been so foolish as not to realise that a beam is stronger when placed on edge? The answer is that slightly curved logs could be halved for joists and rafters as

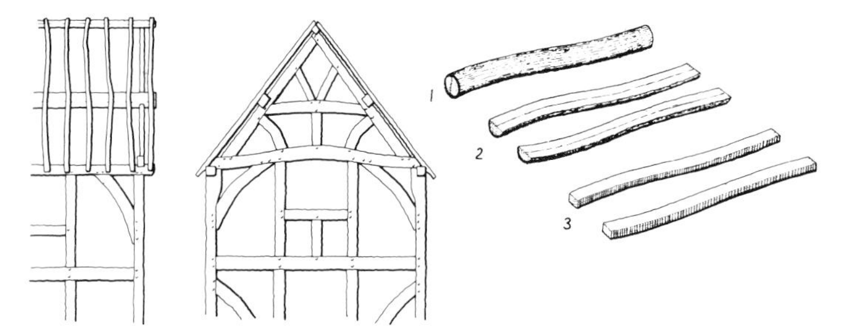

Fig. 14. Left: the use of curved timber. Right: conversion of a log into a pair of rafters or floor joists.

long as the straight surface formed by splitting or sawing – the heart face – was used as the surface to which the floorboards (or roofing battens) were attached; this necessarily means placing the timbers flat rather than on edge.

As is well known, oak was preferred for building, and it was usually used within a year or two of felling. Of the other timbers found, elm is the most common; a few instances of the use of sweet chestnut have been confirmed; and there are a number of examples known of the use of ash, black poplar and other hardwoods.

Evidence is continually coming to light in documents and buildings of the rebuilding and reusing of timber-framed buildings. Sometimes a whole house or barn was dismantled, repaired, and rebuilt, but more often individual beams and frames were reused in a lower-grade building or position. It is fairly easy to identify the original use of such timbers from the joints and marks. Reused timbers are often worth close examination because, however lowly their reuse, they may be the oldest material in the building. Many householders believe that their beams are old ships' timbers, but in few cases yet examined does this seem to be even possible, let alone likely. The reasons for this extraordinarily widespread and persistent belief are still a mystery, but it may derive partly from the former use of the term 'ship's timber' to describe a certain quality of timber – much as the term 'marine ply' is used today.

Timbers of particular character were used in successive periods of timber framing. The earliest surviving buildings contain multiple members which are long and slender, in a style known as 'uniform scantling' carpentry. Later medieval buildings are characterised by heavier timbers and particularly by the use of curved timber in crucks and tie beams as well as in formal designs of arch bracing. In post-medieval buildings the timbers are generally straight, short and regular. No one has yet investigated how far these various changes may have been due to changes in the growth of available timber or how far they reflect changes in carpenters' choice and conversion of timber.

Traditional carpenters' tools were very much like those we know today. Medieval lists and illustrations show saws of all sizes, adzes and axes, chisels, planes, hammers and mallets. Instead of the present-day screw auger, a breast auger with a shell bit was used, and the resulting cup-shaped marks can often be seen in the bottom of mortices. Marking was done with a scribing tool (still used by foresters), dividers and chalk or raddle lines (stretched over the beam and snapped). Underneath the picturesque weather-worn surface we now see lies precise and accomplished carpentry.

Infill panels

Traditional timber frames were intended to be exposed to view. There are some exceptions to this rule, particularly in East Anglia, where certain buildings may always have been intended to be plastered or pargeted, but it seems to have been generally the case until timber framing began gradually to give way to other kinds of walling in the seventeenth and eighteenth centuries: stone, cob, brick, or light timber framing clad with weatherboarding, tile hanging, or plaster.

The spaces between the exposed timbers had to be filled and many different kinds of infill panel were used. The most common is known as **wattle and daub** and follows a similar pattern in most parts of the country. Holes (or slots) about 1 inch (25 millimetres) in diameter are drilled in the underside of the member forming the top of the panel, and a groove is cut in the base member. Oak **staves** are prepared, having a pointed top to fit in the upper holes and a chisel bottom end to slide tightly in the groove. These upright staves, usually spaced about 12 to 18 inches (300 to 450 millimetres) apart, form a strong basis around which **wattles** (usually of hazel or cleft oak) are woven, basket-fashion. The wattle panel is then **daubed** on both sides with a mixture of clay, dung and chopped straw, well mixed and as dry as possible (to avoid

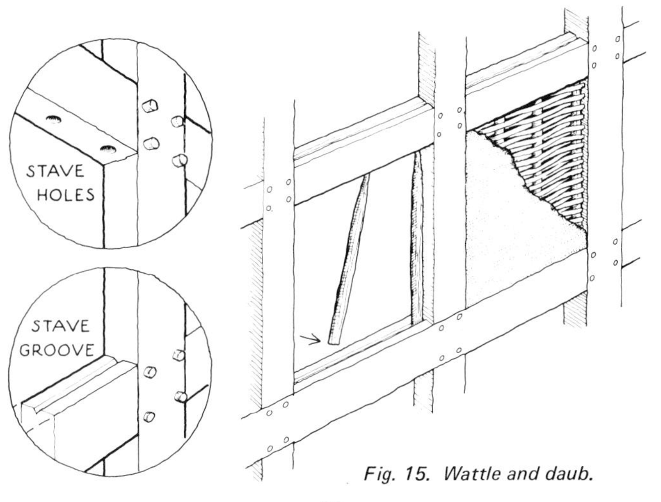

Fig. 15. Wattle and daub.

excessive shrinkage). The completed panel is then limewashed or painted: if properly maintained it will last indefinitely. In East Anglia a slightly different method was used: staves were set horizontally rather than vertically and the ash or hazel wattles were tied to them.

In tall narrow panels there is no room for upright staves and, instead of a woven panel, short lengths of oak lath are wedged horizontally between grooves in the sides of the timbers. In areas where suitable stone exists, thin flat stones are often wedged into the panels instead of laths. Sometimes oak boards are slid into the panels, but this method is found more often in internal screens than in external walls.

Brick is not a good material for infill panels: it is too heavy, tends to hold damp and is a poor insulator against heat loss. Despite this, in the east and south-east it was used quite early instead of wattle and daub, at least from the sixteenth century. In other areas wattle and daub continued in common use until the late eighteenth century. Many wattle and daub panels have subsequently been replaced with brick and many timber-framed buildings have been completely covered or encased with stucco, tile-hanging, weatherboarding or brickwork, usually to their detriment. However, when buildings are altered or demolished it is often possible to examine the timbers for clues as to the original infill material.

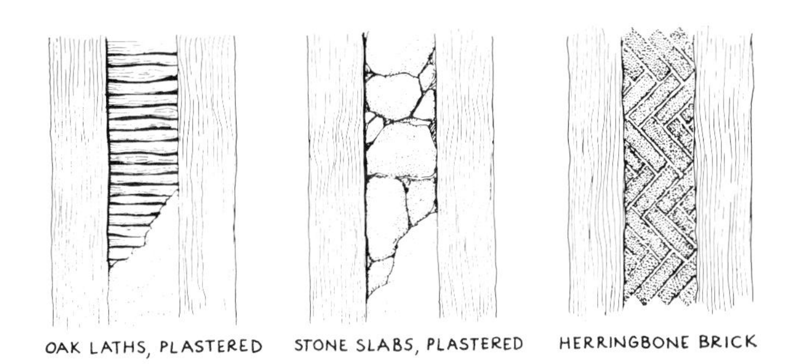

Fig. 16. Infill panels in tall narrow panels (close studding).

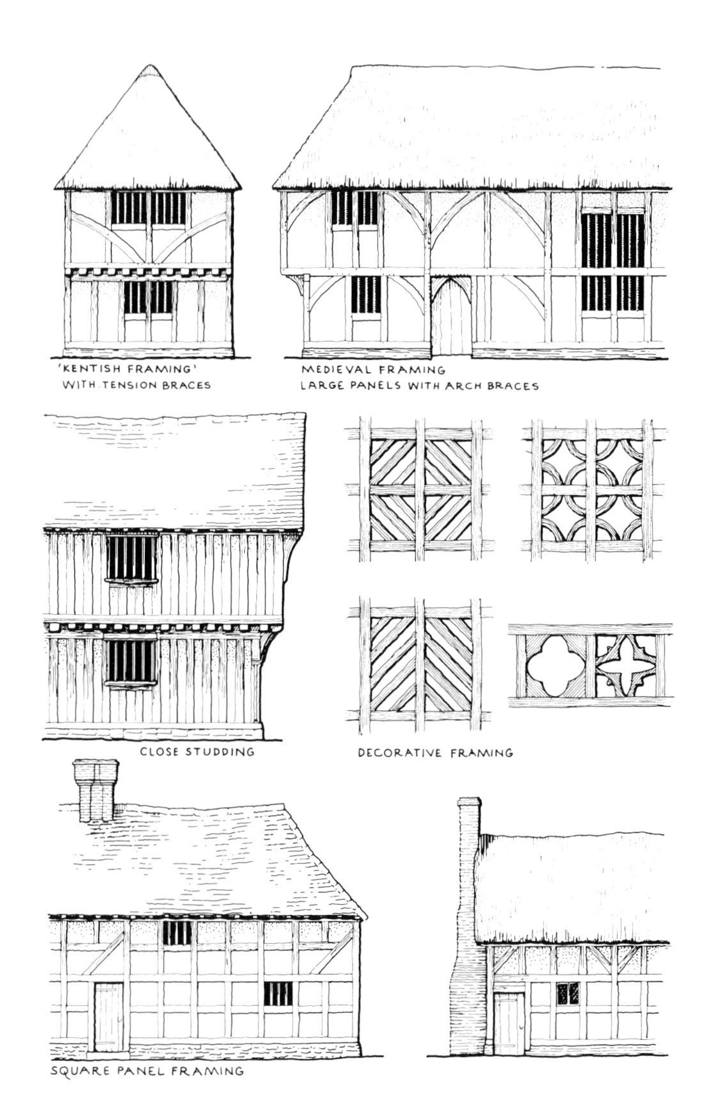

Fig. 17. Patterns of wall framing.

3. Elements

Wall framing

Perhaps the most attractive feature of timber-framed buildings is the great variety of patterns made by the timbers and panels in the wall frames. Nowadays these are customarily emphasised with black and white painting, but there is no evidence that blackening of timbers was widespread before the nineteenth century. In most cases timbers were probably left unpainted, weathering to a natural silver grey: where paint was used to pick out the pattern, earth red or ochre is as likely to have been used as black. Panels were probably not usually stark white but a soft ochre from the impure lime coating or pink from the admixture of sand. Timber buildings which are red and yellow instead of black and white can still be seen in France, Germany and Scandinavia.

The walls of timber-framed buildings tend to let in water through holes in the timbers and gaps around the edges of the panels. The simplest and most effective solution to this is to apply an overall coat of limewash to the wall every year or two, and it is likely that many rural cottages were simply white rather than black and white. Nevertheless it is obvious that in most buildings a part, at least, of the framing was intended to make a pattern.

Most medieval buildings have walls framed in very large panels. Sometimes there are no intermediate rails or studs in the basic spaces formed by the main posts, wall plates, girdings and sills, or just a small number of studs forming large upright rectangular panels. Their most characteristic feature, however, is the use of large curved diagonal braces: in most of Britain these are **arch braces** (rising from post to plate), but in East Anglia and the south-east **tension braces** are more common (running down from post to beam). One particular pattern, having a window centrally placed between two tension braces, is known as **Kentish framing** because of its common occurrence in that county, but it is also often seen on the upper storeys of town buildings in almost all parts of England.

The main types of framing which characterised late- and post-medieval buildings will be discussed in more detail below, but may be summarised here. **Close studding** is universal in East Anglia and may have originated there, but by the middle of the fifteenth century it was in common use in towns and buildings of high status all over the country. Throughout the sixteenth and much of the seventeenth centuries it was used as much as possible by gentry and farmers in most areas: it was

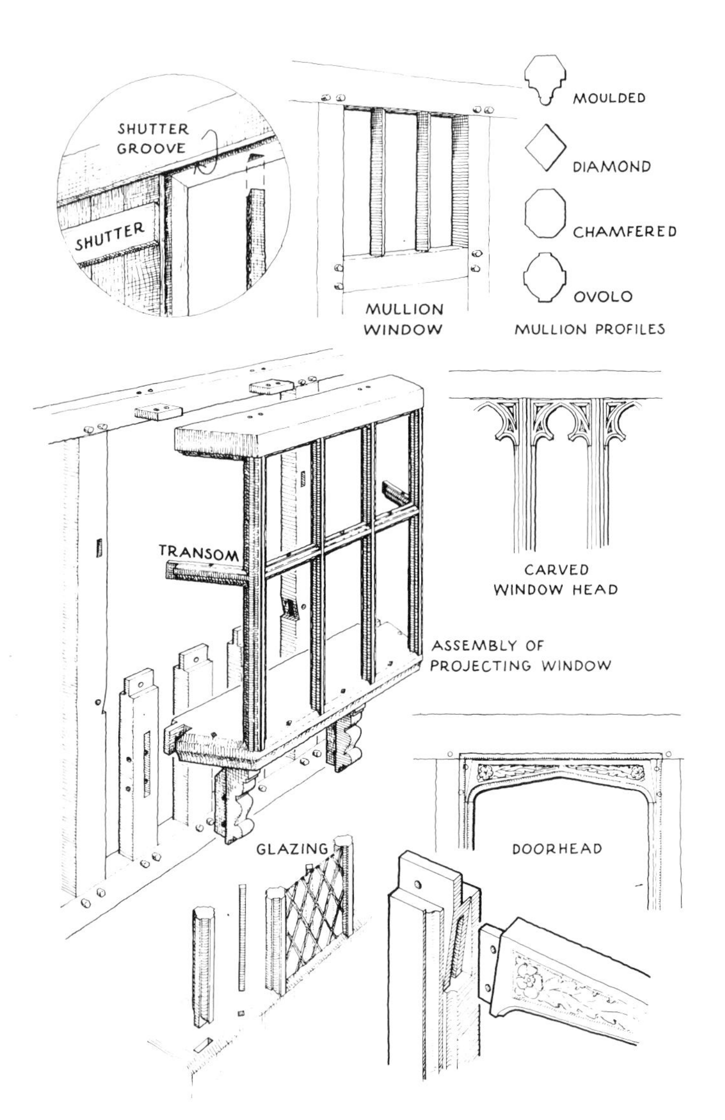

Fig. 18. Window and door details.

expensive in timber and in carpenters' time and was a sign of prosperity and status. Its poor relation was **square framing**, which became common in the late fifteenth century and was used for all the less important applications: back walls, internal partitions, cottages and barns. Love of ornament led to another use of timber as a status symbol in **decorative framing**, in which square panels were filled with a variety of patterns: star, fleur de lys, herring bone, quadrant and others. Plain square panels are common in the south (Surrey, Sussex, Hampshire and Berkshire) as well as the highland zone (the midlands and the north) but neither square framing nor its decorative cousin are found in eastern England.

Windows and doorways

Nowadays we are accustomed to thinking of windows as having frames which are constructed separately from the walls of buildings, usually purchased ready-made from specialist joiners. Luckily for researchers, however, this was not the case in traditional timber buildings, since window openings and mullions were incorporated directly as part of the frame and can therefore be traced from the mortices which remain, even when – as often happens – the original mullions have been removed and a new window inserted. Before the late sixteenth century very few houses had glass: windows were completely open, but the worst of the weather could be kept out with internal wooden **shutters**. These shutters seem usually to have been sliding (vertically or horizontally) rather than hinged, and the grooves in which they slid can often still be seen in the timbers adjacent to the window. Within the window opening itself were vertical **mullions**, spaced perhaps 7 or 8 inches (175-200 millimetres) apart and tenoned top and bottom into the frame. In many cases a **window head** was also incorporated, a horizontal member tenoned at each side into the jambs of the window and elaborately carved to present the appearance of Gothic tracery.

After the late sixteenth or early seventeenth century window openings were designed to be glazed with leaded lights, and they became more elaborate with richer mouldings and the introduction of horizontal **transoms** to divide the lights into smaller panels. Where windows were in line with the wall frame they were still jointed in as part of the frame, but increasing use was also made of **projecting** windows of various kinds. In some cases the projection was slight, with a separate window frame pegged on to the outside of the wall frame; in others it was large enough for a light to be incorporated in the return between the window and the wall frame.

Doorways were treated similarly, the jambs and head being incorporated as part of the main wall frame of the house. As with windows, carved doorheads were used and mouldings carried down the outside of the jambs. The mortices for the doorhead can often still be detected after a doorway has been altered. External doors in houses always opened inwards, and the doors would be hung on hook and band hinges to close flat against the inside of the frame and head rather than into a rebate.

Upper floors

Upper floors consisted of common joists supported by main beams in much the same way as roofs consisted of closely spaced common rafters supported by main roof trusses. Medieval floor joists were usually large, typically perhaps 8 by 6 inches (200 by 150 millimetres) in section, and laid flat rather than on edge. During the sixteenth century there was a gradual change to the use of smaller joists, perhaps 5 by 4 inches (125 by 100 millimetres), placed on edge.

Joists may be placed to run either across or along a building. At an outside wall or a cross frame they may be supported either by a **girding beam** in the frame or by a **ledge** pegged on to the inside face of the frame. If the joists are shorter than the full width or length of a bay there will also be a **bridging beam** to reduce the span. Various combinations of these methods of support are found. Sometimes the joists simply rest on top of the girding beam or ledge (as shown in fig. 19)

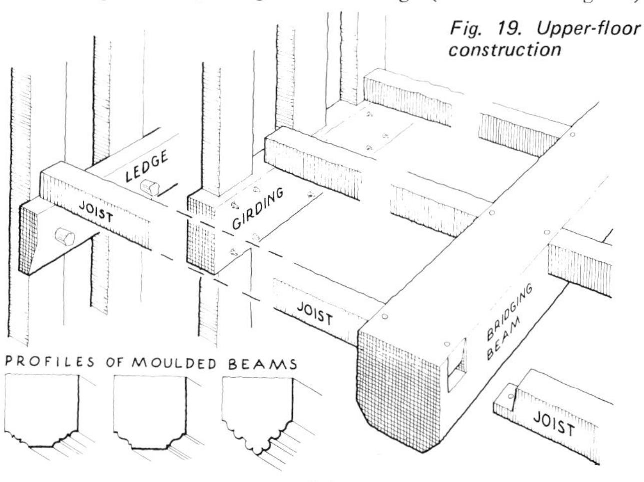

Fig. 19. Upper-floor construction

or they may be jointed so that their upper surface is level with that of the support. Usually the floor boards are nailed across the joists, as in modern practice, but occasionally they are laid in rebates in the edges of the joists so that the boards and joists run parallel and the top surface of the joists becomes part of the floor surface.

Bridging beams within a bay are usually tenoned into the girding beams of the side or cross frames, often with a double tenon for extra strength. They are the largest timbers in the building, typically 1 foot (300 millimetres) square, and have a chamfer or mouldings. In high-quality buildings, particularly from the late sixteenth and seventeenth centuries, there may be two parallel bridging beams in each bay, or a number of beams arranged in a square pattern to give a 'coffered' ceiling.

Plaster ceilings did not become common until the seventeenth century. Before that, floor joists – and the boards above – were usually left exposed. The existence of chamfers or mouldings on the joists gives positive evidence of this when a later plaster ceiling has been applied – unless, of course, the joists originally came from another building.

Chimneys, smoke bays and smoke hoods

Although brick and stone chimneys became a feature of farmhouses in the second half of the sixteenth century, they were not generally built in cottages until rather later. In the transition from open fires to masonry chimneys various timber structures were used to keep smoke from reaching every corner of the house or cottage. Sometimes a short bay, called a **smoke bay**, was sealed off from the upper part of the rest of the house as an escape for smoke. Such smoke bays usually had an area divided off on one or both sides to allow a passageway, an entrance lobby or a staircase. Very often later brick chimneys have been built inside them. An alternative to the smoke bay was a **smoke hood**, a timber-framed enclosure supported on the mantel beam of the fireplace and tapering to an outlet or chimney on the roof. They presented a considerable fire hazard, even though the inside surface would have been plastered (the plaster being mixed with cow dung) to protect the timbers. The hearth under the smoke hood or smoke bay would have heated the hall or main room.

If a timber-framed house was intended to contain a brick or stone chimney the frame was usually designed to allow space for it. It is therefore often possible to judge whether a chimney is original or inserted by examining the frame around it, together with the floor joists and beams which often rest on part of the fireplace structure. If a beam – a tie beam, for instance – disappears into the brickwork of a stack, it is likely that the chimney has been inserted. A chimney which was designed to be inside the house usually has a short bay all to itself, but it is easy to confuse this with a chimney which has been inserted into a pre-existing smoke bay or a medieval cross-passage. On an outside wall the framing was omitted or trimmed to leave an opening into which the stack could be built.

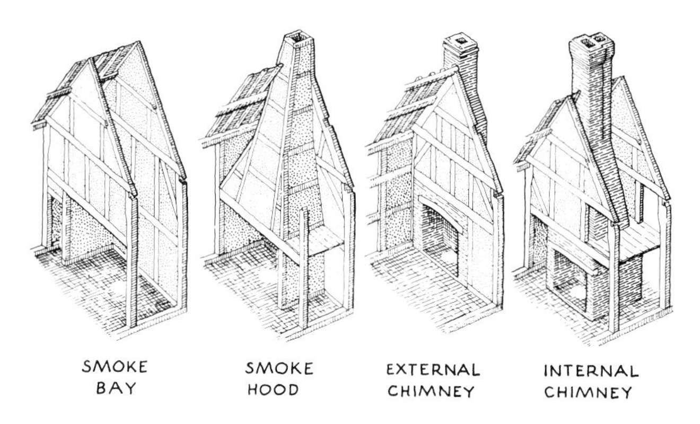

Fig. 20. The development of chimneys.

Wings, turrets and outshots

By far the majority of timber-framed buildings consisted originally of a single rectangle in plan, but there are several ways in which the simple rectangular frame can be extended into wings or turrets, or its depth increased by a rear outshot. In a building consisting of a hall and end cross-wings for example, the structure of each part will often be found to be entirely self-contained: each cross-wing and the hall would stand independently, having its own main posts, wall plates and sill beams. There is obviously a certain amount of waste from duplication in this arrangement and it often points to the various parts of the building having been built – or replaced – at different times: for example, a thirteenth-century cruck hall might have a solar cross-wing added in the fourteenth century, the hall being rebuilt in the fifteenth century and the solar in the sixteenth century. However, to frame a hall and cross-wing so that the end of the hall and the wall of the wing share the same structural members was a rather complicated job, particularly if the two parts of the building had different roof spans or different eaves levels. The course usually adopted was to tenon the hall wall plates into the wall plates of the wing, but there seems not to have been a standard solution for the problem.

Similar observations may be made about the building of small turrets attached to the main building, intended either for a stair or as a two-storey porch. They were often constructed as complete miniature frames, with main posts, tie beams and roof trusses. It is usually possible to decide whether or not they are an original part of the house by discovering whether the wall frame of the house was designed to include doorways giving access out to the turret at upper-floor levels.

Accommodation at the rear of small houses was provided in the form of an **outshot** under a lean-to or cat-slide roof, particularly after the mid sixteenth century. The house would be framed up as normal, with the outshot structure attached to it in much the same way as aisles were attached to the arcade of aisled buildings. Because of this it is often hard to tell whether an outshot is an original part of the building or was added at a later date: one deciding factor is that the outshot was often – but by no means always – used as the site for the staircase. The reverse case is often found in which an original outshot has been removed or rebuilt as a larger extension, and the evidence for this would be a series of mortices in the outside of the principal posts about 6 feet (1.8 metres) from ground level, into which the short tie beams of the outshot would have been tenoned. Outshots were used as service rooms – perhaps a dairy, pantry or scullery.

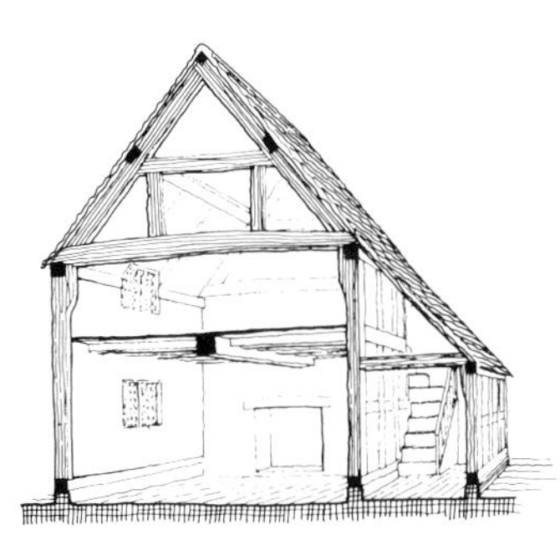

Fig. 21. House with staircase in outshot.

Fig. 22. A medieval open hall.

4. Building types

Open halls

Many enthusiasts would argue that English timber framing reached its zenith in the construction of medieval open halls. In the second half of the sixteenth century, when open halls rapidly went out of fashion, carpenters had to satisfy increasing demands for highly ornate external walls at the expense of the clear and detailed articulation of internal space, which had been the characteristic of medieval timber framing. Certainly every researcher looks forward to the excitement of discovering the heavy, soot-encrusted timbers of the roof of an open hall hidden away in the attic of a much altered farmhouse.

Until the middle of the sixteenth century most people lived in a space which was open to the roof, heated by a fire burning on a hearth built on the floor. For many poor families living in small cottages with low walls built of semi-permanent materials, this was their only living space. Few, if any, such buildings have survived, but some have been excavated by archaeologists working on the sites of deserted medieval villages. The medieval houses which still exist, and which in some areas are quite numerous, belong to a higher level of society. Most of these centred on an open hall (fig. 22).

The main features of medieval halls were widespread and to that extent typical, despite certain differences between the various regions. These differences resulted from uneven socio-economic development as well as from the persistence of deeper cultural traditions, but in all cases the hall was essentially a *formal* space in which the arrangement of various elements of the plan had a clear significance.

Halls were entered directly through a door in the side wall or, more usually, from a **passage** running between a pair of opposed doorways. On one side of the passage would be doorways leading to the **service** accommodation at the lower end of the house: the buttery (for beer), the pantry (for bread), and sometimes an external kitchen. On the other side of the passage there was usually a **screen** dividing it from the hall (hence screens passage). Entering the hall from the screens passage the visitor would see first the **open fire** on a hearth in or near the middle of the floor; and beyond, at the 'upper' end of the hall, the **high table**, with the owner or lord sitting on a bench against the end wall, possibly on a dais raised slightly above the rest of the floor. Above him might be a moulded beam in the wall, or possibly a coved canopy, and, to one side, the doorway leading through to the **solar** or private apartments. Most of the light in the hall came from tall unglazed windows on either side of the high table. In the gloom of the roof would be visible the central open truss of the hall dividing the roof into two bays, all heavily blackened by soot from the open fire.

The hall was evidently a 'hierarchical' space, mirroring the hierarchies of society: the servants at the lower, service, end; the cross-passage entrance; the screen; the 'lower bay' of the hall; the fire; the 'upper bay' of the hall, with the high table; and the private apartments entered from the upper end. Not all halls had all these elements of course, but the *order* was almost invariable, and the timber-framed structures gave expression to it. Briefly, the possible combinations were as follows (fig. 23).

1. An open hall of one or two bays. The cross-passage was sometimes within the hall and sometimes within the service end, and in many cases the upper floor of the service end extended over the passage. In later halls an upper floor may extend over half the space.
2. A bay or cross-wing at the lower end of the house, with service apartments below and a chamber above. Sometimes the service bay may also have been open to the roof.
3. A bay or cross-wing at the upper end of the house. The first floor generally formed the solar chamber; little is known of the use of the ground-floor space except that it was often inferior in status as well as position.

Fig. 23. Forms of medieval houses. Left: single range. Centre: cross-wing at upper end. Right: cross-wings at both ends.

In many cases a hall may have had storeyed accommodation at only one end, usually the service end, and it is common for the hall and ends to have been built – or rebuilt – at different dates.

Structurally, the central open truss of the hall is usually the most interesting and attractive feature of the building, even though in the smoky gloom it must often have been hard to see clearly. The form it took depended on the system of con-

1. Little Moreton Hall, Cheshire. Late sixteenth-century ornamental framing.

2. Paycocke's, Coggeshall, Essex. Early sixteenth-century close studding.

3. String of Horses, Frankwell, Shrewsbury, Salop. Late sixteenth-century close studding. The square-framed end gable (left) would originally have been an internal wall. Now reconstructed at Avoncroft Museum of Buildings, Worcestershire.

4. Moot Hall, Fordwich, Kent. Close studding and jetty carried round all four walls of a public building.

5. Crookhorn Farm, Southwater, Sussex. The cross-wing (right) displays typical medieval framing in large panels. The hall range (left) is a sixteenth-century rebuilding, originally containing a smoke bay.

6. Moat Farm, Longdon, Worcestershire. Close studding, square framing and ornamental framing combined in a typical west midland pattern.

7. Gibbons Mansion, Wyle Cop, Shrewsbury, Salop. Close studding, jetty and projecting windows, late sixteenth century.

8. *Bayleaf Farmhouse, Open Air Museum, Singleton, Sussex. The jetty in its simplest form, and unglazed mullion windows, of the fifteenth century.*

9. A simple crown-post roof in a medieval farmhouse. The wattle and daub partition is a later insertion, as is the ceiling which obscures the tie beam.

10. Roof truss with trenched purlins in a Cheshire barn, probably seventeenth century.

11. Netherhale, St Nicholas at Wade, Kent. A fifteenth-century moulded crown post, the tie beam obscured by an inserted ceiling.

12. The Bromsgrove House, Avoncroft Museum of Buildings. A simple fifteenth-century roof with trenched purlins and curved windbraces.

13. Detail of cruck construction in a medieval Herefordshire barn.

14. *The Barn, East Riddlesden Hall, Keighley, Yorkshire. A late medieval Yorkshire aisled barn with king-post roof.*

15. *Warehouse, Howard Street, Shrewsbury, Salop. King-post roof construction in the nineteenth century.*

16. A fourteenth-century carved windbrace from the Guesten Hall Roof, Avoncroft Museum of Buildings.

17. A nineteenth-century brace from a cowshed near Petworth, Sussex.

18. A wall frame laid out prior to erection.

19. A timber frame during assembly at the Open Air Museum, Singleton.

20. A fifteenth-century Wealden house with typical later modifications: a brick chimney, an upper floor in the hall, the ground floor framing replaced with brick, and a rear wing added.

21. The same house after reconstruction at the Open Air Museum, Singleton.

22. A lap-dovetail joint between tie beam and wall plate, from a nineteenth-century building. The precise cutting of the joint contrasts with the poor quality of the timber. The carpenter's assembly mark can be seen cut into the base of the dovetail.

23. The same joint re-assembled.

24. Wattle and daub panels. The finished coat has been applied only to the panel under the window.

25. A panel of split oak pales woven around oak staves in a Herefordshire barn.

26. A building of 1572 which was dismantled and reconstructed in the nineteenth century. The porch and lower windows were added at the whim of the architect but are of the wrong period. Victorian reconstructions such as this have given timber framing an undeserved reputation for 'fakery'.

27. The same building (The Hop Pole, Bromsgrove, Worcestershire) in its original state, pictured by M. Habershon in the early nineteenth century.

struction used in the hall, which varied from area to area (fig. 24). Most examples have arched braces below the tie beam or collar of the truss and in the grandest examples the shape of a true arch is formed. It is interesting that in several cruck-built halls there is also a horizontal beam below, which apparently destroys the lofty arched effect. Such a beam undoubtedly had a structural function in preventing the crucks from spreading and may on occasion have been used for other purposes, suspending a cooking pot over the fire for example, but perhaps it also had a symbolic function in articulating the upper and lower bays of the hall.

Another curious feature associated with central open trusses is difficult to explain. Despite their name, they are rarely placed centrally: one bay of the hall is almost always longer than the other, sometimes by only a foot or two, sometimes by considerably more. Sometimes the upper bay was longer, sometimes the lower bay. There can be no structural reason for this: medieval carpenters were quite capable of dividing a structure into equal bays. We can only assume that the placing of the open truss had some symbolic or practical significance in relation to the use of the space below.

Mention has already been made of the distinction between the upper and lower faces of frames, and of the fact that their positions bear a traditional relationship with the plan of the building. In this case the upper face of the central truss in an

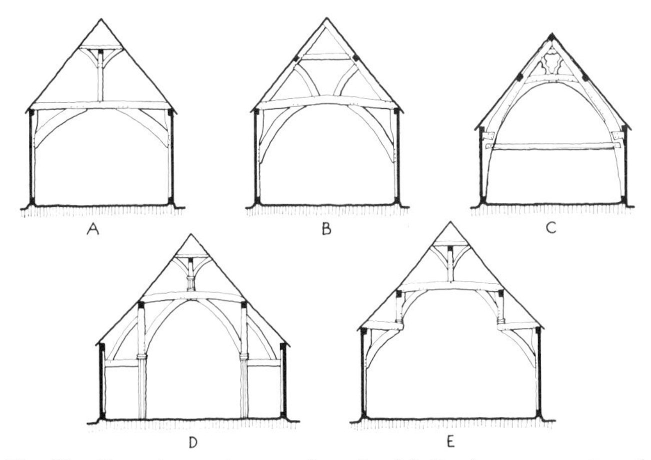

Fig. 24. Central open trusses of medieval halls. A: crown-post roof. B: purlin roof. C: cruck frame. D: aisled. E: hammer beam.

open hall was generally placed to face towards the dais – the upper end of the hall. The difference between the upper and lower face would not be noticeable from ground level, but carpenters working within a strong craft tradition in a strongly ordered society evidently felt it was right – in the nature of things – that the best face of the frame should face their masters at the best end of the hall. This is the kind of intimate and highly evolved connection between craft, structure and space which justifies the claim that building carpentry reached here its highest level of excellence.

Post-medieval houses

It is useful to remember that the greater the age of a building the more likely it is to have been built by someone occupying a high position in society. Medieval halls were generally either manor houses or, in some areas, the homes of prosperous yeomen or industrialists, and the very earliest – from the twelfth and thirteenth centuries – belonged to the most important lords. From the sixteenth century onwards an increasing number of peasant farmhouses survive, and the smallest cottages date from the late seventeenth century and after. Part of the reason for this is, obviously, that the best buildings survive longest, but it seems also that in the late sixteenth and seventeenth centuries the houses of the majority of the population started to be built in much more permanent materials – timber frame, stone or brick – where previously they had been of poor construction and customarily rebuilt completely every generation or so. This change, which laid the foundation for much of today's rural scene, has been called the Great Rebuilding of rural England.

It is very rare for any surviving building to have only one room: the minimum accommodation is generally one heated and one unheated room, in which case the latter would presumably have been where food was stored. Tiny cottages of this kind do survive, either with a heated room downstairs and a chamber above or with two small rooms under one roof and no upper storey. However, most post-medieval houses are of two storeys and have either a two-room or a three-room ground plan. Such houses can best be understood by noting the position of the entrance and the chimney and analysing the relationship between the two (see fig 25). The main distinction is usually between those which have a cross-passage and those in which the entrance leads either directly into the main room or into a lobby against the chimney. It is an important distinction because the two types of plan follow a regional distribution similar to that of different forms of construction in the highland and lowland zones of England and Wales.

Cross-passages, which were an almost universal feature of medieval halls, remained an important post-medieval feature in the houses of peasant farmers in parts of the highland zone in the seventeenth and eighteenth centuries, especially in the south, west and north-west, the limestone belt and the Pennines. In most cases the stone, brick or timber chimney was placed between the cross-passage and the hall (the main ground-floor room) forming a 'chimney backing on entry' arrangement. There are also a number of cross-passage houses with the chimney placed elsewhere – either on a side wall or at the upper end of the hall; these are often found to be of high (or relatively high) status.

The second main class of plan, characteristic of lowland zone houses, is the 'lobby entrance' type, so called because the front door leads into a small lobby against the side of the chimney. Many medieval open halls were converted into lobby entrance houses by inserting an upper floor in the hall and building a chimney in the cross-passage, demonstrating the

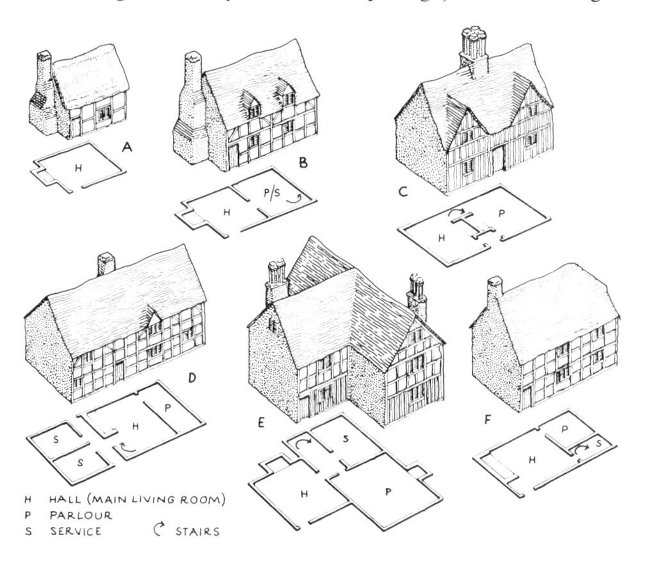

Fig. 25. Post-medieval houses. A: one-room plan. B: two-room plan. C: lobby entrance plan. D: cross-passage plan. E: T-shaped plan. F: end lobby entrance plan, with two inner rooms.

continuity of plan traditions in medieval and post-medieval houses. Such houses became extremely common in East Anglia and the south-east in the seventeenth century. Contemporary examples are known also in parts of the midlands and Wales and during the eighteenth century houses and cottages with this kind of plan were built in most parts of Britain.

A third type of plan is important in the north and west midlands: the T-shaped farmhouse, consisting of a two-bay wing across the end of a hall range of one or two bays. The positions of entry and chimney vary: sometimes they form a lobby entrance at the junction of the hall with the wing and sometimes they follow a cross-passage arrangement at the lower end of the hall. The use of the two bays of the solar wing varies less: one forms the parlour while the other is occupied by the staircase and a small pantry.

Of these three groups of plans the cross-passage type is least often found timber-framed – simply because the areas in which it is most common are those where the usual building materials are stone or cob. Examples will also be found, particularly in the midlands, of timber-framed farmhouses larger than the types outlined above – houses with two wings, for example, or with double-pile plans – mainly dating from the late seventeenth century. Timber buildings were attractive, impressive and much loved, and only the forces of fashion and scarcity of materials brought the tradition to an end for the construction of substantial houses.

Urban details

Until the seventeenth century the majority of buildings in towns and cities in England were timber-framed, except in those areas in which easily available stone was always the normal building material. Vestiges of the scene this presented still exist in certain streets in, for example, York, Stratford-upon-Avon and Lavenham, but timber framing could not survive the combined attack of disastrous fires and the new fashion for brick. During the seventeenth and eighteenth centuries the appearance of our old towns changed completely, as it is now being changed again in modern redevelopment.

Technically timber framing in towns followed much the same patterns as in the country, but with a few important differences. Buildings in towns (particularly those of high status) tended to make use of certain forms of construction which were not used in the surrounding countryside: crown-post roofs and recessed-bay halls are the two main examples. Similarly, the display of wealth was more common in town than countryside, so timber-framed buildings in towns tended to have ornate fronts and generally to lead fashion in decorative treatment. Perhaps the most obvious urban feature was the use of oversailing upper storeys (known as *jetties*), which were extremely common in fifteenth- and sixteenth-century town buildings but only sporadically used in the countryside. Also, the competition for space in the commercially valuable city centres led to the use of condensed and modified forms of the plans of rural houses and to the erection of precarious-looking 'skyscraper' buildings of three or four storeys.

Party walls make an interesting study. In some medieval cities there were regulations which demanded that a stone wall of a certain thickness should be built between separate properties in order to provide a fire break, and many of these walls are still standing, even though the timber-framed structures between them were later rebuilt. In other places, however timber-framed walls were allowed as party walls and it seems that in many cases a single thickness of framing infilled with wattle and daub panels was felt to provide sufficient separation.

Many – but by no means all – medieval town houses consisted of a shop facing the street, with a chamber above and an open hall behind. In such cases the structure could be arranged to present a gable to the street, the roof running back over

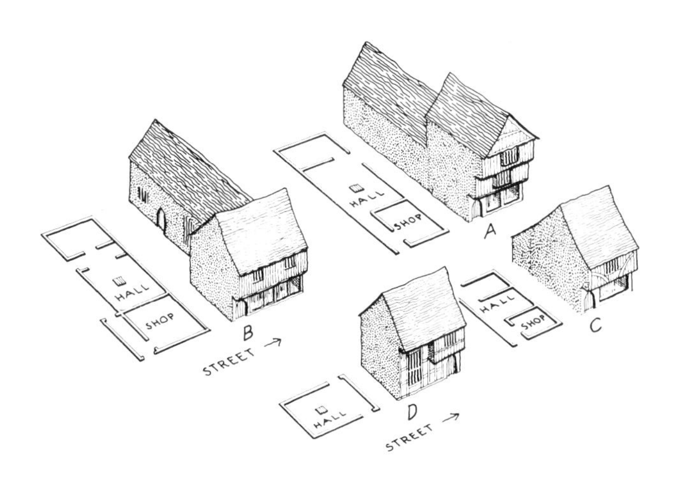

Fig. 26. Medieval town buildings.

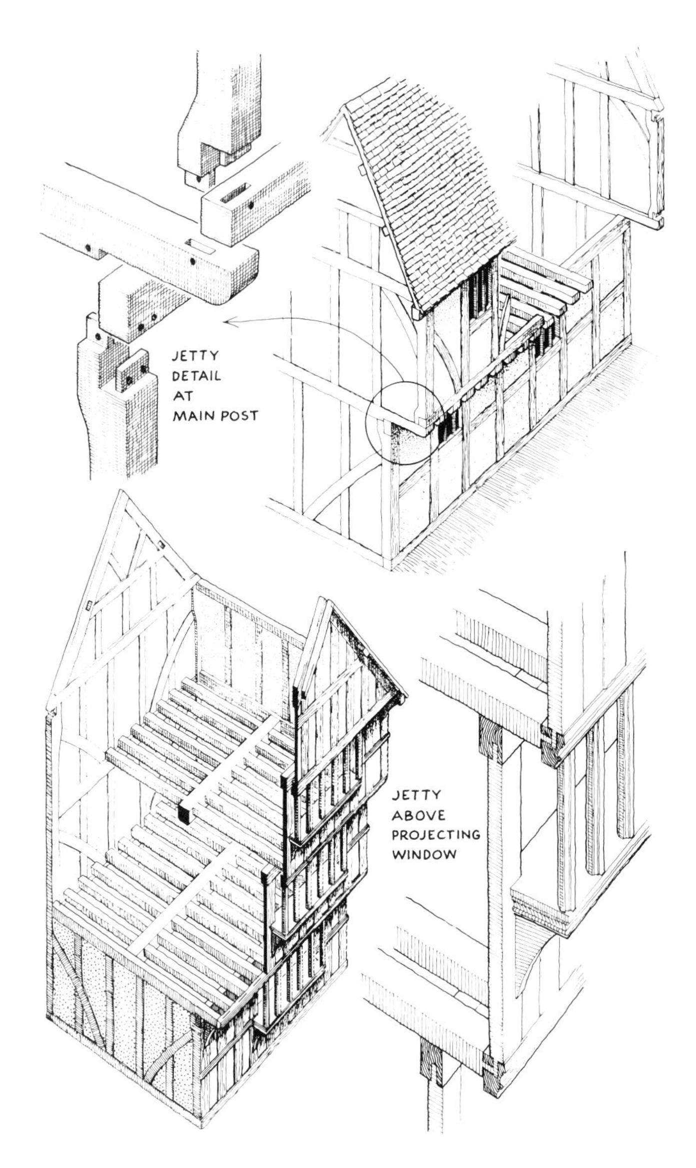

Fig. 27. Jetties and projecting windows.

the hall and other buildings (fig. 26A). It was also possible for the front range and the hall behind to be covered with two parallel roofs (fig.28) or for the front roof to be parallel to the street and the hall range behind to be roofed at right-angles to it (fig.26B). In the fascinating row of buildings in Church Street, Tewkesbury, the accommodation was so compressed that the hall and chamber of each unit were contained under a single-span roof parallel to the street (fig. 26C). The roof could thus continue without a break over the entire row of twenty-two units, probably built as a speculative development for letting to artisans. Other examples of comparable developments of long rows of identical small units are known elsewhere, such as in York and Coventry, mainly with recessed-bay halls and jettied chambers facing the street (fig. 26D).

Jetties (fig. 27) are a most important feature of timber framing in towns, as well as being used for superior farmhouses in some areas. Basically, a jetty consists of an upper wall projecting or jutting beyond the wall below, forming an overhang. The wall of each storey is constructed as a separate frame, the upper one being supported on the ends of the floor joists which rest on, and project beyond, the wall frame of the storey below. In the simplest case the ends of the floor joists are exposed to view between the two storeys, the lowest beam of the upper wall frame simply resting on top of them. In

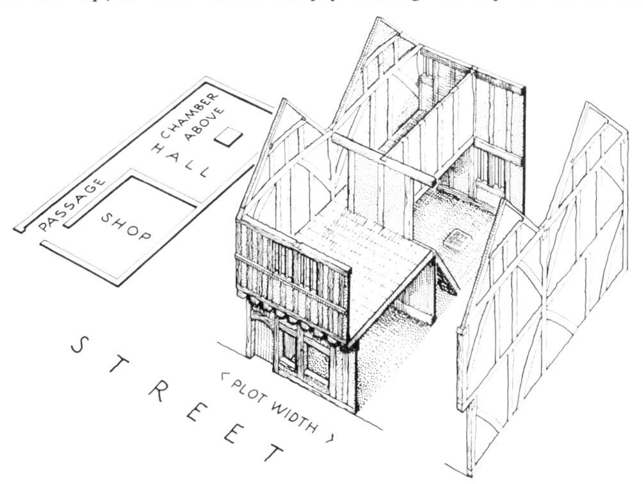

Fig. 28. Medieval town building (based on example reconstructed in Spon Street, Coventry.)

more sophisticated examples the same beam is morticed on to the front of the joists, thereby concealing them, and is usually moulded. In all examples the main posts and beams are jointed together.

The projection of jetties varied from a few inches to two feet or more: 1 foot 9 inches to 2 feet (525-600) millimetres) was perhaps the usual standard in the fifteenth century. In buildings of high quality it was quite common for the mullions of projecting windows in the storey below to be tenoned into the soffit of the jetty beam above, or for a plastered coving to be constructed below the projection. Unfortunately centuries of maintenance and repair have taken a heavy toll of such features, but their previous existence may often be detected from the mortices and peg holes used in their construction. When enough money was available, jetties offered opportunities for carving and decoration on beams, posts and brackets.

Jetties were undoubtedly an architectural symbol of wealth and status. A man who could afford only one jetty on his house would want it to face the main public thoroughfare, but more wealthy patrons could sometimes afford a jetty facing a side alley or a courtyard as well. Important detached buildings such as market halls were jettied on all four sides for maximum effect. Jetties can be constructed either on the gable end or side wall of a building, but the joists supporting the upper storey must always run outwards under the jetty. For this reason, when two jetties join at right angles a diagonal beam known as a *dragon beam* was introduced into the floor structure and the corner post below took the form of a *dragon post* with a head enlarged outwards in both directions and used as an opportunity for a great display of carved decoration.

As well as being valued as a feature of the architecture of timber-framed buildings, jetties also gave certain practical advantages which may have been equally valued. Probably the most important was the ten to twenty per cent increase in floor space that a single 2 foot (600 millimetre) projection could give to each storey. In addition they provided passers-by with a measure of protection from rain. It has also been argued that the main-span deflection (sagging) of the floor joists is reduced by the introduction of a reverse bending moment where they oversail the lower storey, but this effect is probably insignificant and there is no evidence that carpenters were aware of it. Overhanging upper storeys are a common feature of other European carpentry traditions but constructed in such a way that none of these functional considerations apply.

Finally, mention must be made of the **continuous-jetty house**, a very widespread building form of the sixteenth and seventeenth centuries. The plan of such houses usually follows one of the post-medieval patterns described above, but the whole of one side of the building is jettied to the street or to the village green. It should not be assumed, however, that the existence of a continuous jetty necessarily means that the building to which it belongs is a house. All buildings of any importance – guildhalls, market halls, schools and houses – were jettied, and a careful examination of the interior arrangement must be made in order to discover the purpose of a particular example.

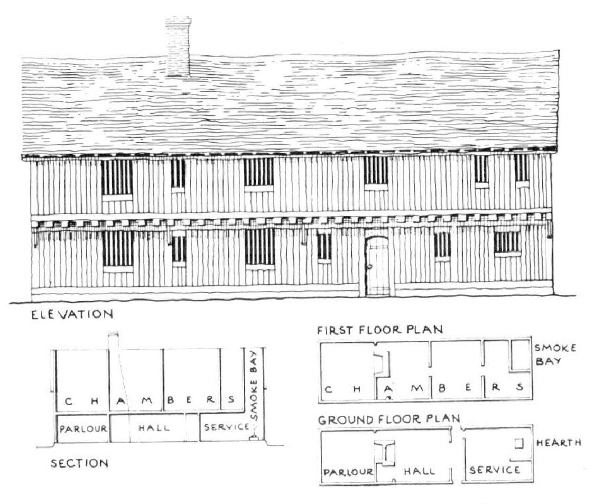

Fig. 29. A continuous-jetty house.

Farm buildings

Farm buildings offer by far the best opportunities to study timber framing. Most houses have been altered at various times, and the details of construction are often obscured by plaster inside and black paint outside, but in barns the only obstacle to vision is usually bales of straw – and even these can afford a useful means of climbing to inspect the roof timbers at close quarters (always, of course, with the owner's permission). The construction of barns follows the prevailing patterns in each area but in a simple form and without any decorative detail.

Over the whole of Britain **barns** are remarkably uniform in plan. In the simplest (and most common) case the barn is three bays long, with two storage bays flanking a central bay in which a threshing floor lies between the large entrances on either side. One doorway is usually larger than the other, to allow for a heavily loaded harvest wagon to enter. The crop – wheat, barley, oats or rye – was threshed with flails on the stone, earth or timber threshing floor during the winter months, the two doorways providing a through draught for winnowing. In some cases one end of the barn is longer than the other, allowing room for threshed straw, which is more bulky than the unthreshed crop tied in neat sheaves. Larger barns are simply enlargements of this basic plan. It is rare to find an original upper floor in a barn, although often a floor has been inserted within the last hundred years: traditionally the crop was stored right up to the roof (just as straw bales are nowadays).

The infill or cladding materials in barns differ from those in houses. Crops in store need ventilation and many timber-framed barns probably originally had wattle panels left undaubed: in most surviving examples of original panel infill the weaving is made of wide (3-4 inch, 75-100 millimetre) cleft oak pales rather than the hazel or narrow oak laths used in wattle and daub. Most later timber barns were originally weatherboarded, and many have since been infilled with brick.

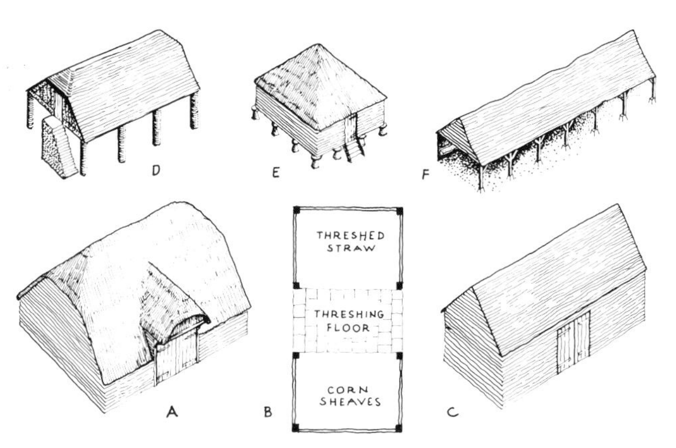

Fig. 30. Farm buildings. A: aisled barn. B: plan of three-bay threshing barn. C: non-aisled barn. D: granary above cart shelter. E: granary on staddle stones. F: open-fronted shelter shed.

Timber-framed **granaries** are often found, although they were rarely built earlier than the late eighteenth century; before then grain was customarily stored in an upper floor of the farmhouse. Granaries in lowland England are usually single-storey timber-framed buildings, raised on stone staddles or brick piers to protect the grain from damp and rodents. The other main type, found in the south-east as well as the midlands, consists of a first-floor granary with an open space below used as a shelter for carts and wagons. Granaries of both types vary greatly in size. There is usually a single door (with secure lock) and a single window, and the interior is completely plastered and lime-washed.

Accommodation for cattle was of two kinds: the open-fronted **shelter shed** or hovel, in which fatstock kept loose in the foldyard could shelter and feed; and enclosed **cow-houses** or byres, in which cows could be stalled in the winter and where milking took place. Both kinds of building demonstrate simple local timber building methods, but few examples will be found to have survived from before the late eighteenth century. The timber feeding racks and troughs and, in byres, the stall divisions are also of interest.

Haylofts are sometimes found above byres but more often above **stables** for working horses. Timber-framed stables are found in some areas, but even in those built of brick or stone the roof and upper floor construction are of interest. They are seldom built with a full two storeys and often contain instances of 'interrupted tie beam' or 'upper cruck' construction in order to allow the whole of the upper space to be used for hay.

Farm buildings are working buildings and tend to be completely rebuilt more often than houses. It is not uncommon to find a farmhouse to be the oldest building on a farmstead, but early examples of farm buildings are rare and deserve close attention.

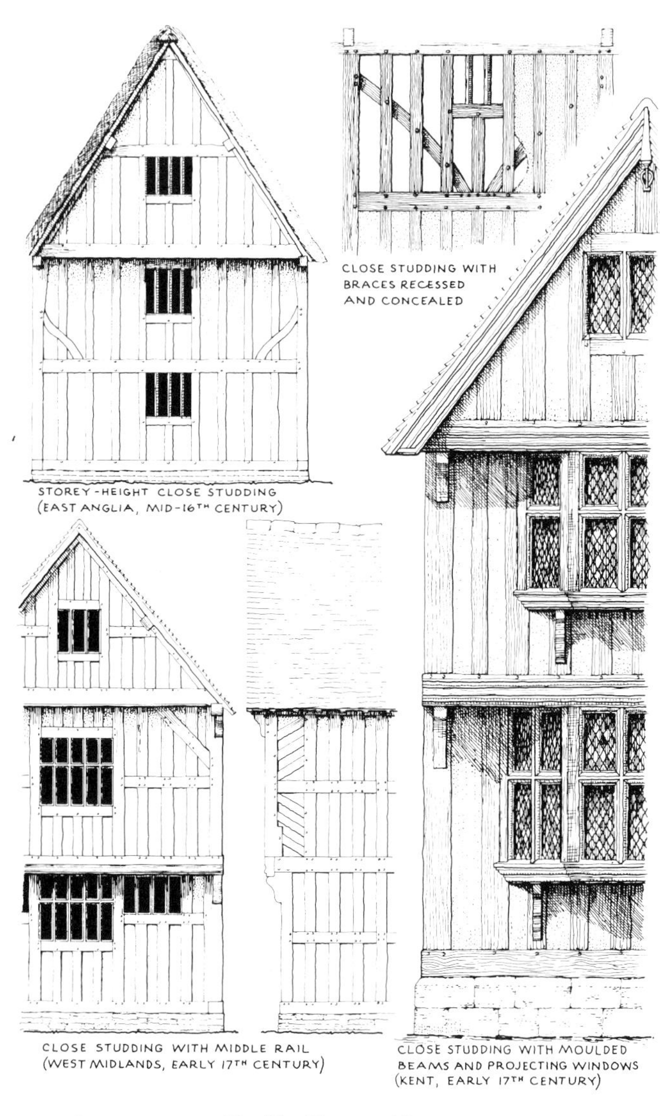

Fig. 31. Close studding.

5. Lowland zone: the east and south-east

Introduction

The local character of English timber framing is not easy to define. Some features and details vary from county to county, or even from parish to parish; others can be seen as characteristic of regions such as East Anglia or the west midlands; and a surprising number define a strongly national tradition, suggesting that for carpenters the thirty miles of the English Channel presented a greater gulf than the three hundred miles from the north to the south of the country.

The following three sections of the book are intended to give an indication of the regional character of the tradition, and of the way in which some features – close studding, for instance – spread from their place of origin. The three regions do not have precise geographical boundaries, but they are convenient headings under which characteristic features may be grouped.

Close studding

Close studding is the main distinguishing feature of the school of carpentry centred in East Anglia, walls being composed of storey-height studs set fairly close together. The earliest surviving East Anglian timber walls, which are probably early thirteenth-century in date, are close studded, and the style persisted unchallenged in the area until at least the seventeenth century. The spacing varies, sometimes becoming as much as 2 feet (600 millimetres) or more, probably owing to lack of timber – or money – rather than to any change in the desired visual effect. Frames need diagonal bracing for stability, but this would spoil the close-studded appearance, so braces were therefore often halved across the inside faces of the studs and concealed by the plaster panels externally. Undoubtedly the pattern would have been enhanced both internally and externally by the application of coloured paints and limewash.

Close studding is concentrated in East Anglia, but it is also common in the rest of lowland England. It seems to have become fashionable in the early fifteenth century and was used for two centuries, particularly in buildings of high social status. Such buildings included not only the houses of the gentry and of prosperous farmers, but also public buildings such as guildhalls, market halls and churches. It was particularly common in towns: a foreign visitor to London in 1497 remarked on the impressive close studding and other towns must have been similar.

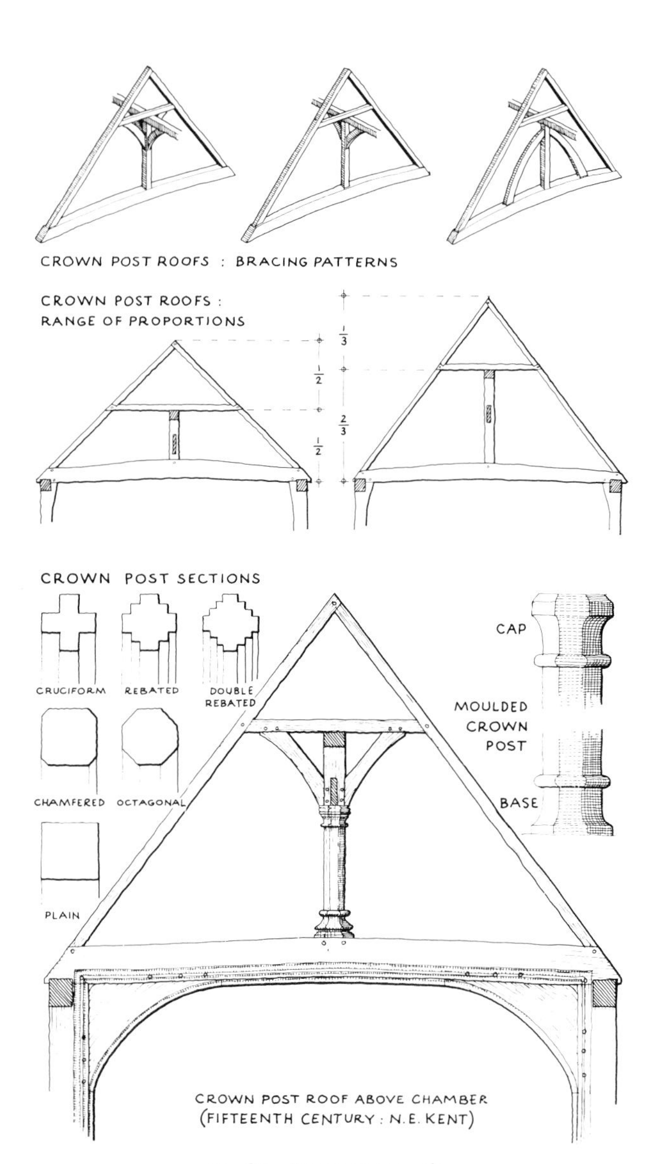

Fig. 32. Crown-post roofs.

This applies also to the rest of Britain: wherever there was timber framing, close studding became the most desirable – and expensive – pattern from the sixteenth century onwards. In the west midlands it is possible to make a rough guess at the status of a seventeenth-century farmer by finding out how much close studding he could afford. Sometimes if the cost was too high, studding was faked with planks or paint, just as timber framing has often been faked in this century. There is a difference, however, between highland and lowland patterns: lowland studs are always a full storey height, whereas the highland studs are usually divided by a middle rail. Cruck buildings with close-studded walls are extremely rare: there seems to have been an almost unbridgeable cultural gap between the two styles of carpentry.

Crown-post and clasped-purlin roofs

Crown-post roofs were described briefly (page 8) as one of the three main types of roof truss. Although very early crown-post roofs are known, they appear to have developed from an earlier, simpler form of roof in which there were no longitudinal members at all, each pair of rafters being joined by a collar and held in place by the battens and roof covering. Obviously such roofs ('single' rafter roofs) would be liable to collapse – like a pack of cards – and it is thought that crown plates, supported on crown posts, may have been introduced partly to provide some longitudinal stiffening and partly to help support the collars – and hence the load of the roof.

The design of crown-post roof trusses varies mainly in the height of the crown post and the pattern of diagonal bracing between crown post, crown plate and collar (fig 32). Unfortunately none of these features has undisputed dating significance. Crown posts in the central open trusses of halls or solar chambers were often given a cap or base with decorative mouldings and cut to octagonal or other sections as a decorative feature, but in less exposed positions and humbler buildings they are usually square and slender.

Crown posts are found in great numbers in the whole of eastern and south-eastern lowland England, but hardly at all elsewhere. Where examples exist in other areas they are in medieval buildings of high status, usually in towns, suggesting that crown-post construction may have been associated with the ruling or merchant class. After the early sixteenth century they rapidly became scarce and their place was taken by **clasped-purlin roofs** (fig. 33).

Clasped purlins of a kind can be seen in some of the earliest surviving roofs (for instance the Wheat Barn, at Cressing

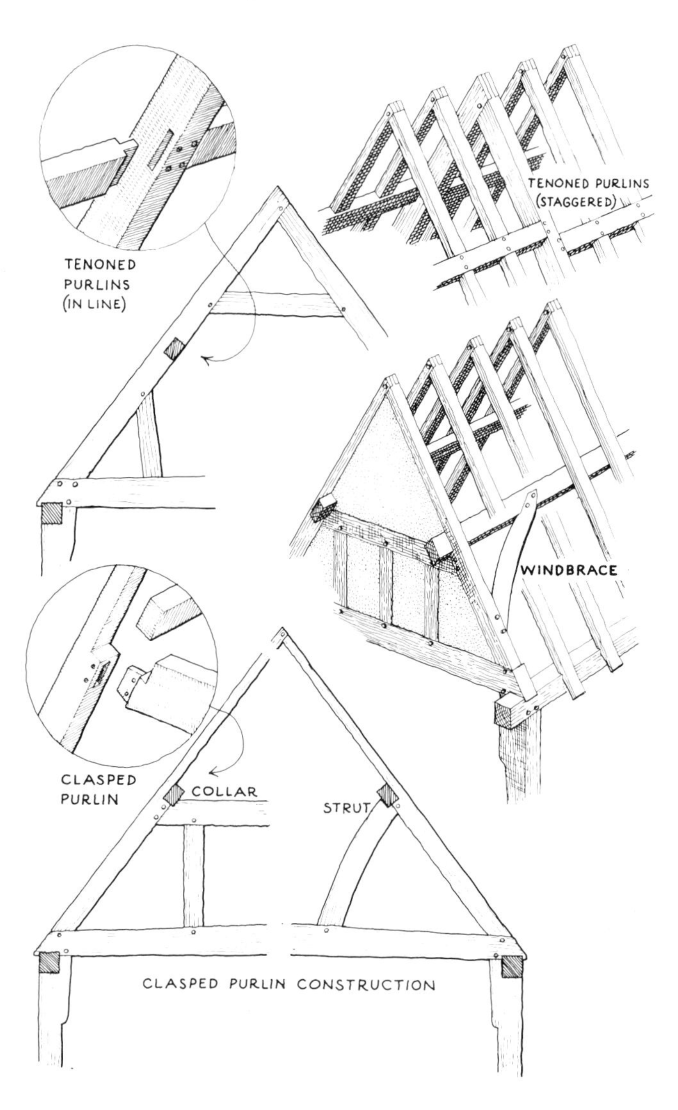

Fig. 33. Clasped-purlin and tenoned-purlin roofs.

Temple, Essex, about 1250), and scattered examples were built throughout the whole of the medieval period – during which crown-post roofs were dominant. After about 1500, however, the position is reversed and clasped purlins became the standard, almost universal, form in lowland England until the nineteenth century. They can also be found in highland areas – cruck country – in the fifteenth and sixteenth centuries but later yielded to roofs with heavy principals as the dominant form. A characteristic of clasped-purlin roofs is that the principal rafters (those which are tenoned into the tie beams at the bay divisions) have their upper surface in line with the common rafters and are in most cases only slightly larger in section. This contrasts with roofs having trenched purlins supported on heavy principals, in which the principal rafters are much larger than the common rafters and not in line with them.

Another method of fixing purlins is to tenon them into the principal rafter. Tenoned purlins are found in most parts of lowland England, used as an alternative to clasped purlins, and in other areas they are sometimes a feature of high-quality medieval carpentry. Two variations of tenoned purlin construction are shown in fig. 33: staggered purlins with tenoned rafters seem to be the later form.

Wealden houses

In highland England the characteristic (but by no means universal) house of the medieval rural peasant was a cruck house, open to the roof in the centre but often with upper floors at each end. In lowland areas, and particularly the central weald of East Sussex and west Kent, a different type of house predominated in the medieval period, its superiority reflecting the more settled agricultural prosperity of the area and the influence of London. This is the so-called **Wealden house**, an arrangement of an open hall and jettied end chamber or chambers under a single roof giving a distinctive external appearance. Although most common in the Weald, it is also quite widely distributed elsewhere in East Anglia and the south-east.

Jetties were undoubtedly an architectural symbol of status, whatever their other advantages may have been, and in Wealden houses they are an integral part of the design. Upper chambers at each end of the building, with their floors jettied out, flank a central hall which is open to the roof and therefore has no jetty. This gives rise to the characteristic external feature, the recessing of the hall behind the jettied upper chambers, the roof being carried in an unbroken line from one end of the building to the other.

Because the roof is carried past the recessed hall (on a 'flying' wall plate) a Wealden house is built in a single range with hipped ends. Inside the roof, however, the structure may imitate a hall with cross-wings: above each chamber there is sometimes a tie beam on the axis of the building (fig. 34), carrying a crown post and giving the effect of a two-bay chamber across the end of the hall – as if in a two-bay solar cross-wing.

Wealden houses were most common in the fifteenth century, but there are also some dating from the sixteenth. The most common form had a two-bay hall and a jettied chamber at each end, but several examples may be found with only one end chamber, and a few examples have only a single-bay hall. Many Wealdens have survived intact and can still be

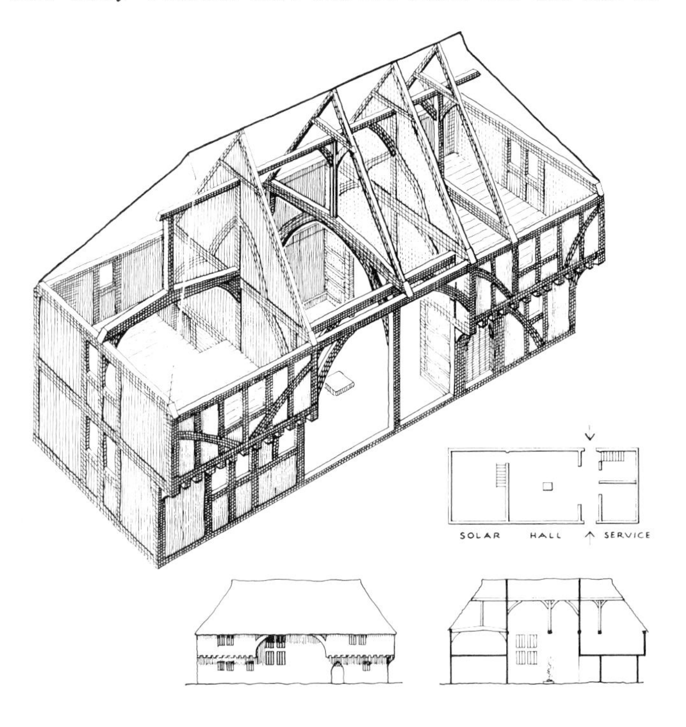

Fig. 34. A Wealden house.

recognised externally by the recessed wall of the hall, despite the upper floor which will have been inserted in the open hall and the brick chimney which replaced the original open fire. These changes were usually made in the late sixteenth or seventeenth century.

Small hall-houses of Wealden type, having a recessed-bay hall, are sometimes found in towns in highland areas: there are examples in Stratford-upon-Avon (Warwickshire), Weobley (Herefordshire), Newark (Nottinghamshire), York and elsewhere. This is another example of forms of building which originated in the south-east being used elsewhere, apparently because they carried connotations of high status – the equivalent, perhaps, of our modern 'executive-style' housing developments. There are very few instances of the reverse process – the use of highland forms in lowland areas.

Wealden halls are not the only type of medieval hall found in lowland England. Houses roughly equivalent to a Wealden in size, but having jettied ends rather than sides, will lack the characteristic recessed hall of the Wealden. Larger houses often have cross-wings at one or both ends. It is the architectural quality of Wealdens, however, which best expresses the quality of life of a prosperous age.

Aisled barns and halls

The use of aisled construction represents one of the fundamental schools of carpentry in England, and many examples of aisled barns can be found in the lowland counties of Kent, Sussex, Surrey, Hampshire, Berkshire, Hertfordshire, Essex, Buckinghamshire, Suffolk, Norfolk and Cambridgeshire. Within these counties the distribution may be more localised: in Sussex, for example, they are found mainly on the Downs and the coastal plain rather than in the Weald, reflecting the need for larger barns following improved grain yields in the late seventeenth and eighteenth centuries. Although such barns clearly served farming needs, the use of aisled construction must also be seen as a continuation of the regional style of carpentry. In other areas of rich farming, such as Herefordshire, aisled barns are completely unknown. In Kent and East Anglia there seems to have been an unbroken tradition of aisled barns from medieval times right through to, in some cases, the nineteenth century.

The construction of aisled barns varies in several respects in different areas and periods (fig. 35). The main roof over the nave tends to follow the same local patterns as non-aisled buildings of similar date, medieval crown-post trusses giving way to clasped or tenoned purlins. The basic construction of the aisle consists, in most examples, of a main wall post with a jowled head jointed to the wall plate and to a short 'aisle tie beam' which is tenoned in turn into the main arcade post inside the barn. There is sometimes also an 'aisle principal rafter' or a strut supporting a purlin over the aisle, but more usually the common rafters span unaided from arcade plate to outside wall plate. The main variations in aisle construction occur in the pattern of bracing: until the seventeenth century

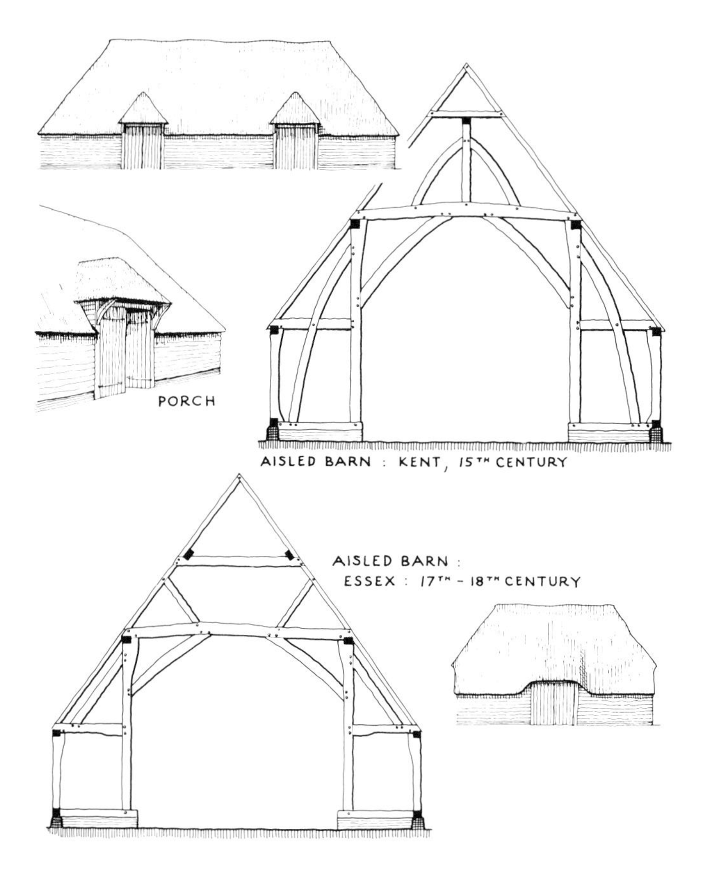

Fig. 35. Aisled barns.

the most common method was to incorporate a long curving member (sometimes known as a 'passing brace' or 'passing shore' running from the sole plate in the aisle to the outside of the arcade post and halved over the aisle tie. Subsequently, bracing of the more straightforward type was used, triangulating between the aisle tie or sole plate and the wall post or arcade post.

Aisled barns also vary in plan: the most common type has an aisle at each side, but in some cases only one side is aisled and in others the aisle returns around one or both ends, the construction being similar to that of side aisles. All aisled barns must have at least one porch or raised doorway, allowing access for loaded wagons. These can also take several forms, sometimes hipped, sometimes gabled, sometimes projecting beyond the wall of the barn. With all these points of difference it is clear that the external shape of aisled barns can vary quite widely, but the interior construction is always instantly recognisable.

Aisled barns are extremely numerous and easy to find in lowland England. Aisled halls, on the other hand, are rare, early in date (almost all before 1400) and therefore have received far more attention than their descendants in the farmyard. Further discussion of aisled halls and aisle-derivative structures (see page 10) is beyond the scope of this book: the interested reader should consult chapter 1 in *English Vernacular Houses* by Eric Mercer and the references given there.

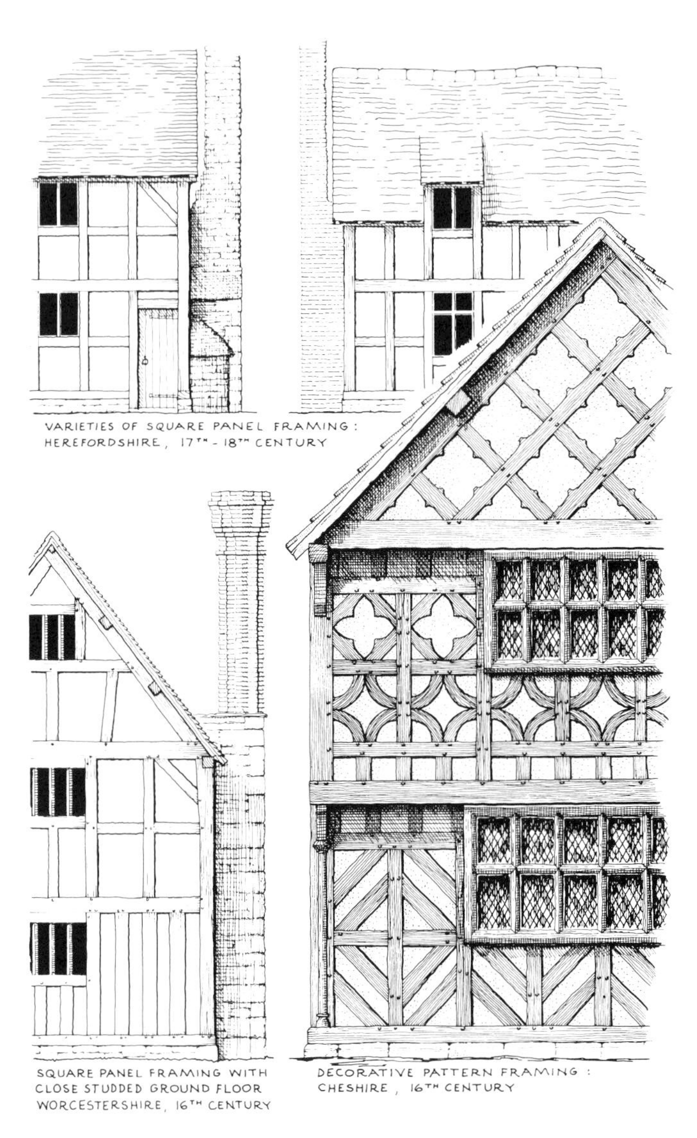

Fig. 36. Square panel and decorative framing.

6. Highland zone: the midlands, south and south-west

Square panels and decorative framing

A most common sight in the midlands is of farmhouses, cottages and barns with their walls framed in panels about 2 feet 6 inches to 3 feet (750 to 900 millimetres) square, three or four panels high from sill beam to wall plate. Very often the panels are filled with brick, but originally they would usually have been wattle and daub. Most often the whole building is square-panelled, but many houses displayed the higher status of their owner by having close studding or even decorative carving on the main facade. Sometimes only the ground floor walls are close studded, with square panels in the upper storey. A brick or stone chimney will usually be visible attached to an end or side wall, or appearing through the roof.

Square panels appeared first in the west midlands, particularly Gloucestershire, Herefordshire and Salop, in the mid fifteenth century, but during the sixteenth and seventeenth centuries they were used in large numbers of buildings in a wide belt from Cheshire to East Sussex and Kent. They can be seen in good houses as well as in cottages and barns, but close studding or ornamental framing was always preferred if it could be paid for. There was also a tendency in the sixteenth and seventeenth centuries for higher-quality buildings to have rather small square or rectangular panels: the more timber was used, the more impressive was the house.

Decorative framing is based on square panels and is found over much the same area, but with a particular concentration in Lancashire and Cheshire – including such well-known examples as Little Moreton Hall. It became common in the late sixteenth century and remained in use for nearly a hundred years. Elizabethan and Jacobean love of ornamental display, both internal and external, was prodigious. With timber houses painted in 'comely colours', chequerboard patterns in chalk and flint, diaper brickwork, star-sectioned chimneys, stripes, figures and mottoes, our towns and countryside must have presented a riot of colour and ornament quite different from our sober modern taste.

Crucks and later roofs

Crucks are known over very wide areas of England, but they probably reached their most highly developed and impressive form in the west midlands and the Welsh Marches, in medieval halls and barns. The largest known cruck building is

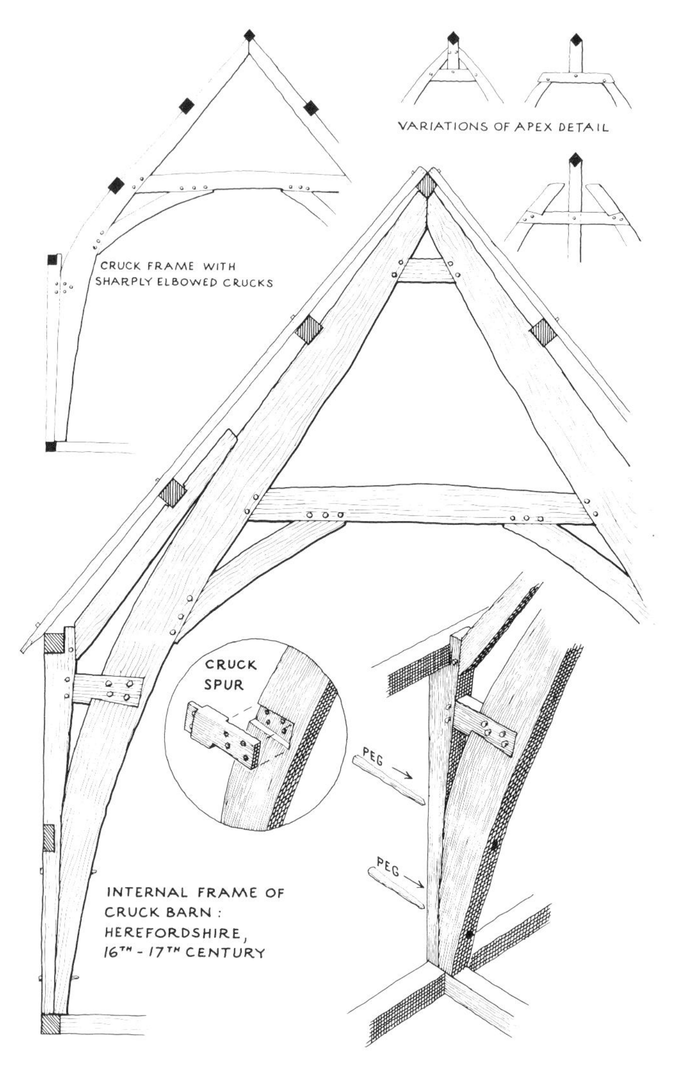

Fig. 37. Cruck construction.

in west Worcestershire (the great barn at Leigh) but the structural details used are just the same as in a multitude of smaller buildings.

The crucks carry the roof load to the ground, and the vertical walls are usually subsidiary frames which depend on the crucks for stability and support. Sometimes there are no wall posts at all, the crucks themselves carrying the wall plates, but more usually there is a separate wall post pegged to the outside of each cruck and firmly linked to it by a **cruck spur** halved to the face of the cruck. In barns the internal cruck frames will all be of similar design, but in hall-houses the design of the various frames will reflect their position and function in the building.

Some crucks are smoothly curved or nearly straight, others are sharply elbowed, the shape depending on the shape of available trees. Generally (but not always) a tree would be sawn in half to give a pair of crucks of identical profile – a technique which was also used for other paired members such as arch braces. Not all crucks reach far enough to support the ridge piece: again this usually depended on the size of timber available. Most crucks are oak, but other timbers were occasionally used, such as a small number of black poplar crucks which have been found in Worcestershire and Herefordshire.

Crucks were used both for barns and for houses, although there are more surviving examples of the latter. The cruck halls of manorial lords and prosperous peasants follow the general outlines described on page 31; the most important halls, however, were of post and truss construction, heavily decorated and sometimes with crown-post roofs – the use of lowland construction as a mark of prestige. At a lower social level cruck buildings were probably universal (or practically so) to judge from references in contemporary documents, but very few examples seem to have survived. These lesser buildings will not have had the full medieval hall plan, their one, two or three bays housing the villein and his family – and probably his animals – in the simplest conditions and with the minimum of accommodation.

In the midlands and south most crucks are probably medieval: during the sixteenth century cruck building dwindled. This was a change in fashion, but it may possibly have been aided by the heavy demand for curved oak for shipbuilding. Crucks were superseded by post and truss buildings in highland England in much the same way as crown plates were superseded by clasped purlins in lowland areas. The roof trusses which succeeded crucks are of the type having heavy principal rafters which support trenched purlins and (usually) a ridge:

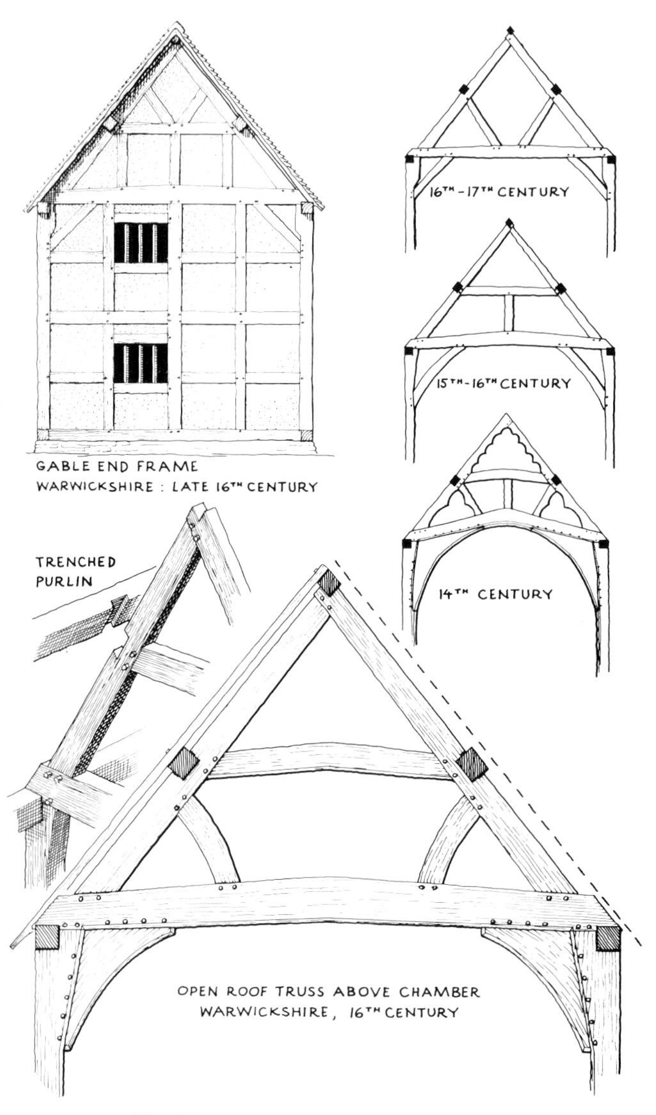

Fig. 38. Roof trusses with trenched purlins.

crucks support the purlins and ridge in the same way and, to that extent, the carpentry tradition was unbroken.

Roof trusses with tie beams and heavy principals do not vary greatly in their design (fig. 38). Tie beams, particularly in medieval buildings, may be cambered upwards – sometimes quite steeply, with the centre of the tie beam forming the point of an arch. Within the truss a collar and/or struts are arranged so as to support the principal rafters where they carry the purlins. Virtually all post and truss buildings used the assembly of post, tie beam and wall plate known as the tie beam lap-dovetail (fig. 9), which was a constant feature of traditional carpentry in England.

Roof trusses without tie beams

Medieval buildings with an upper floor tend to have a full-height second storey, so that the tie beams of the trusses do not impede access from one bay to another, but in buildings of high status intermediate trusses are sometimes found in the centre of a main bay, composed simply of a collar between a pair of principal rafters. Arch braces up to the collar can be used to make an impressive Gothic feature of the truss (fig. 39). In these trusses the principals are simply notched over the wall plates and are not accompanied by a main post in the wall frame. Trusses of this kind can only be used in buildings with side purlins, and no medieval equivalent seems to have been employed in crown-post roofs.

In smaller post-medieval houses it became common to build the main frame a storey and a half high, in which case a tie beam at eaves level would obstruct the upper storey. Three main solutions were adopted. The most straightforward was to omit the centre of the tie beam, the remaining ends being tenoned into upright members rising from the main cross beam (supporting the floor) to the principal rafter or collar beam above. This method (forming what could be called **interrupted tie beam** trusses) still left obstructions at either side and it was commonly used to form a doorway through the truss rather than to open up the whole space. An alternative was to use **upper crucks**. These are like true crucks except that they rise from the first-floor cross beam. In some cases the traditional cruck spur attaching the wall plate to the cruck is replaced by a wrought iron strap, producing a very simple and economical structure (fig. 39).

Both of these methods are virtually restricted to the highland zone, although the use of members similar to upper crucks has been recorded in East Anglia. In clasped purlin roofs another method was sometimes used, known as the

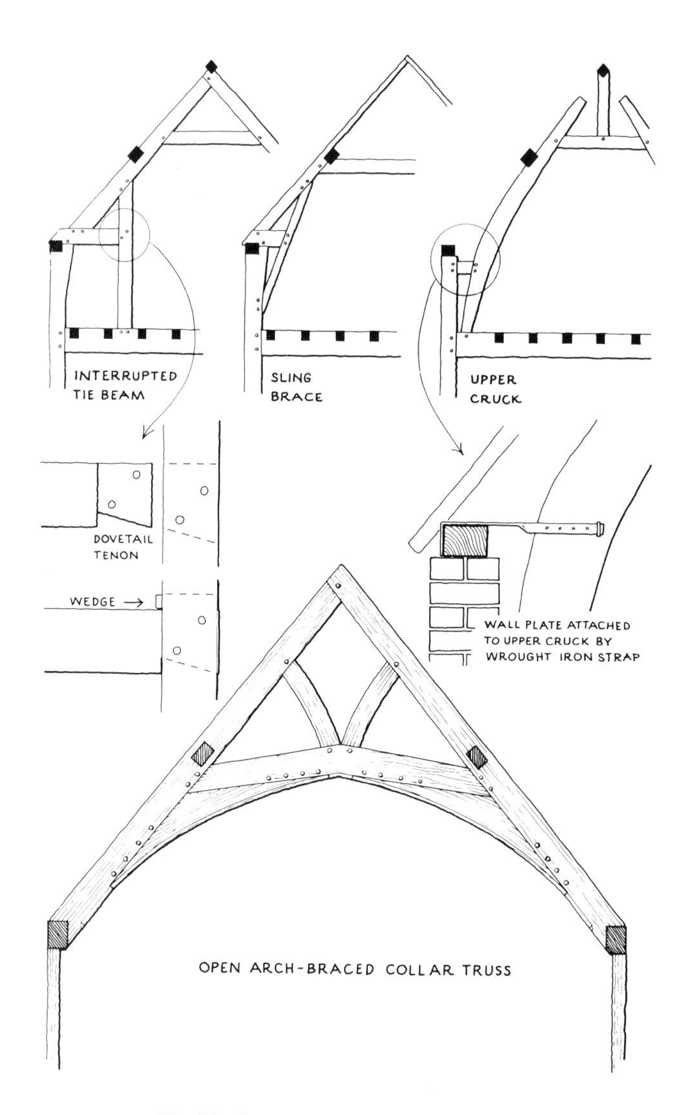

Fig. 39. Roof trusses without tie beams.

sling-brace truss. This is similar to the interrupted tie-beam type except that, instead of a vertical post, a diagonal member is introduced running from the wall post to the rafter; a short section of tie beam completes the necessary triangulation.

All these methods of framing a roof without a tie beam were extensively used in brick buildings and farm buildings as well as in timber-framed houses. They are part of the last chapter of our story, in which timber framing is relegated to use in the floors, roofs and internal partitions of houses and farm buildings.

Jointed crucks in Dorset, Somerset and Devon.

The south-west is not an area for timber-framed buildings; most rural houses have walls of stone or cob. Internally, however, timber was used as richly as in the west midlands and the south-east. Internal partitions, for example, are often **plank and stud** screens, well built and carefully chamfered and shaped. In roof construction some full crucks are found of medieval and later date, built into the cob or stone walls, but the most characteristic form is the **jointed cruck** in which two timbers are jointed together to form a composite member of cruck-like shape. The most usual joint is a long mortice and tenon, but examples are also found with a simple splayed scarf, face-pegged. Some buildings with jointed crucks are certainly medieval, but the technique continued in use during the seventeenth and eighteenth centuries.

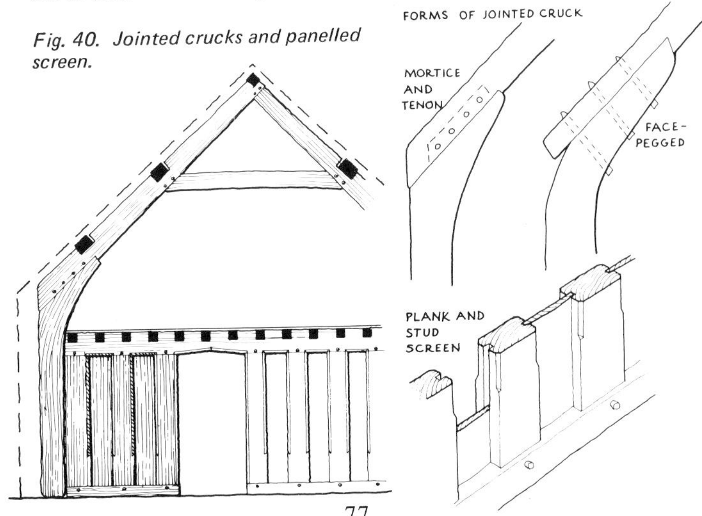

Fig. 40. Jointed crucks and panelled screen.

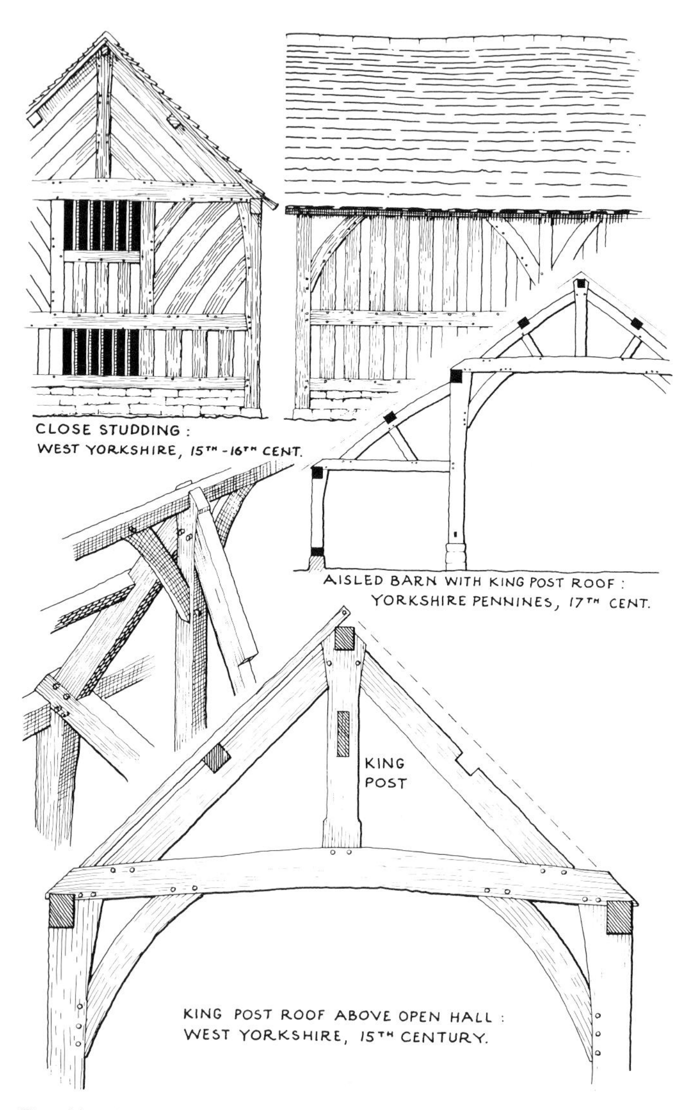

Fig. 41. Close studding, aisled barn and king-post roof in Yorkshire.

7. Highland zone: the north

Northern crucks and king posts

Timber framing in the north of England, particularly Yorkshire, has been called the third school of English carpentry, contrasting with the eastern lowland and western highland styles. Viewed from the other regions the north breaks all the rules. There are king posts, hardly known elsewhere in England until at least the mid seventeenth century. There are crucks, but mostly poor and late rather than the substantial medieval forms surviving in the rest of the highland zone. There are rafter roofs, roofs with clasped side purlins, aisled halls and barns, and close studding – lowland features but with a distinctive northern style.

Northern crucks present quite a different picture from the more highly developed and intricate cruck construction of an earlier period in the south and west midlands, although the two traditions were undoubtedly connected. The most obvious difference is that northern cruck buildings generally have stone external walls – perhaps replacing earlier sod, clay or 'clam-staff and daub' – where midland crucks have timber-framed walls jointed to them. The jointing of northern crucks is rougher and less precise and it is easy and tempting to see in them the possibility of a straightforward descent from early and primitive construction. The crucks themselves are generally rather roughly hewn and jointed together with a single cross beam to form an A-frame (fig. 42). The cross beam is halved across the face of the crucks and projects on each side to support the wall plates. In some cases the purlins rest on the projections of a collar or upper tie beam also halved across the crucks; in others they are supported by the crucks themselves. Examples are also found of open cruck frames, in which the wall plates are tied to the crucks with spurs, as in medieval crucks in the midlands, but of rougher timber and construction. Unless they are embedded in stone walls the crucks rest on pad-stones or 'stylobates'. In spite of the lower standard of construction, northern crucks were erected as described on page 16, by being assembled flat on the ground and then reared into position: small rectangular mortices, which were used as sockets for levers during the rearing, are often found near the base.

In the north crucks are more often found in small barns and hovels than in houses and seldom seem to date from before the seventeenth century. Houses built on crucks will often be found to have been small cottages open to the roof as a single-storey dwelling, perhaps with upper floors or lofts in the unheated ends. Despite being open to the roof these seventeenth- and eighteenth-century cottages would not have had an open hearth: the fire would have been built in an inglenook beneath a stone or timber hood and chimney.

The poor crucks of the north of England have received much attention in published works by three pioneers of the study of traditional buildings, S. O. Addy, James Walton and C.F. Innocent, who all worked in Yorkshire and Lancashire. This historical accident has contributed to a widespread belief that all cruck construction is essentially rough and primitive, a remnant of the practices of our technically incompetent ancestors, even though the earliest surviving cruck buildings are of relatively high quality and not at all primitive. It should be emphasised that most examples of poor crucks date from the seventeenth and eighteenth centuries. They exist side by side with the well-built stone houses of more prosperous farmers. Similar poor cruck dwellings undoubtedly existed in other areas in earlier periods and descriptions of comparable buildings in Warwickshire and Leicestershire have been published.

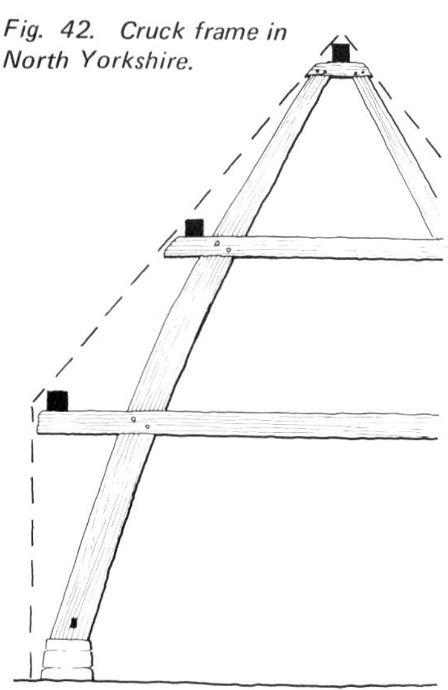

Fig. 42. Cruck frame in North Yorkshire.

Medieval halls in the Pennines of West Yorkshire present some striking contrasts with contemporary buildings in other parts of the country. The roof trusses are of **king-post** construction and are generally sturdy and plain, lacking mouldings or ornamentation. The halls are often **aisled** and, although they were lofty and open to the roof, the fire was built on a hearth under a timber smoke hood rather than in the middle of the floor. The walls are framed with close studding, the studs arranged in two rows of unequal height. The main posts rest on large foundation stones (stylobates) and the sill beams are tenoned into them rather than running beneath them as in the rest of the country (a feature known as the 'interrupted sill') (fig.41). Almost all of these medieval Pennine timber buildings were cased in stone in the seventeenth century and later, and few can be recognised from outside.

In a king-post truss a stout post rises from the tie beam to give direct support to the ridge of the roof. The principal rafters on each side rise from the tie beam and tenon into the sides of the king post; the purlins, generally rather slender in section, are trenched into the principals. Although there are a small number of early king-post trusses in other areas, they are extremely common in the north in medieval and post-medieval buildings. In the eighteenth and nineteenth centuries the form was widely adopted in the rest of the country, perhaps owing to the influence of builders' pattern books.

Aisled buildings in the south-east of England have already been described, but there is also an extraordinary concentration of aisled halls and barns in the southern Yorkshire Pennines. As in the south-east, the surviving examples consist of a small number of medieval halls and a large number of barns, most of which are post-medieval. The existence of the aisled halls may have been related to the rapid expansion of the Halifax cloth industry. The aisled barns are more difficult to explain: farming in the area was based on pasture rather than arable and it is probable that the side aisles of the barns were used for housing cattle rather than corn. A notable feature is that in later barns the aisles are often as wide as the nave and have separate access from outside. The outside walls are usually of stone, but some may originally have been timber.

Aisled timber buildings occur also in the Vale of York, the wide plain lying to the east of the Pennines, but with roofs of common rafters and collars – at first without any form of purlin, later with clasped side purlins. The wall framing generally contains close studding and arch braces, like the Pennine halls but lacking their distinctive arrangement into panels of unequal height. These buildings are obviously comparable with those of East Anglia – only a short sail away to the south – but with heavier wall frames and without crown-post roofs.

Many of the aisled houses in Yorkshire have an aisle on one side only, allowing tall windows to let in light on the non-aisled side. If a partition is inserted into the arcade to divide the nave from the aisle the house becomes similar to those built later with outshots. It seems that in Yorkshire these two forms merged one into the other without a break, whereas in south-eastern England there was a gap of almost two hundred years between the last aisled hall and the common adoption of outshots in the early seventeenth century.

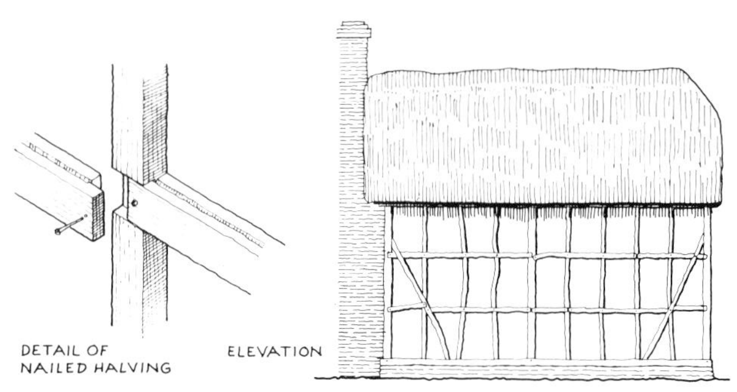

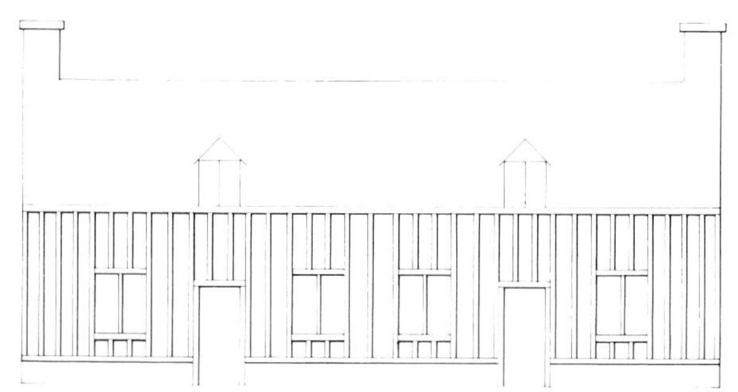

Fig. 43. Eighteenth-century framing.

8. The end of the tradition

Cottages and barns

In timber, as in any other material, to build well costs money and in all periods most people have had to live in poor and insubstantial buildings. None of these buildings has survived from the middle ages, but a number are known which date from the late sixteenth century onwards. Many of them are in Lincolnshire and Lancashire, where a technique of walling known as 'mud and stud' or 'clam-staff and daub' was used in conjunction with minimal timber frames. In these buildings the wall frame typically consists simply of wall plate, sill and main posts, perhaps with some diagonal braces and very light intermediate studs and rails. The walling material is a clay mixture like daub or cob, but applied much more thickly and strengthened by thin upright staves nailed to the frame. Other examples of buildings of similar status and rough timber construction are the cruck cottages of Yorkshire and Cumbria.

As is well known, timber supplies were increasingly in demand during the seventeenth and eighteenth centuries, and house carpenters were in competition with shipbuilders, iron smelters and others for supplies. The best timber was used – or reused – where it was essential, in the roofs and floors of important buildings, leaving the poorer quality to be used as economically as possible in cottages. In some eighteenth-century Worcestershire cottages, for instance, the main posts, wall plates, sills and roof members are still of reasonable size, but the studs and rails within the frame are extremely thin, little more than squared poles, lapped and nailed to each other to form large square panels infilled with wattle and daub in the time-honoured way (fig. 43).

However, it is easy to exaggerate this tendency to use smaller timbers. Many eighteenth- and nineteenth-century farm buildings, mills and other utilitarian structures were built in timber of a quality comparable with that used in the sixteenth and seventeenth centuries, albeit with more elm and softwood than oak. Neither can it be said that standards of craftsmanship declined: what had changed were the demands made of carpenters. Instead of being asked to frame a whole building, they were increasingly required simply to fit a roof or a floor into a brick structure, or to frame up a wall to which external boarding, tiles or plaster would be applied. Timber framing had ceased to be architecture and had become simply an element of structure. Whereas previously carpentry techniques had been handed on from master to apprentice in an essentially local tradition, carpenters in the eighteenth and

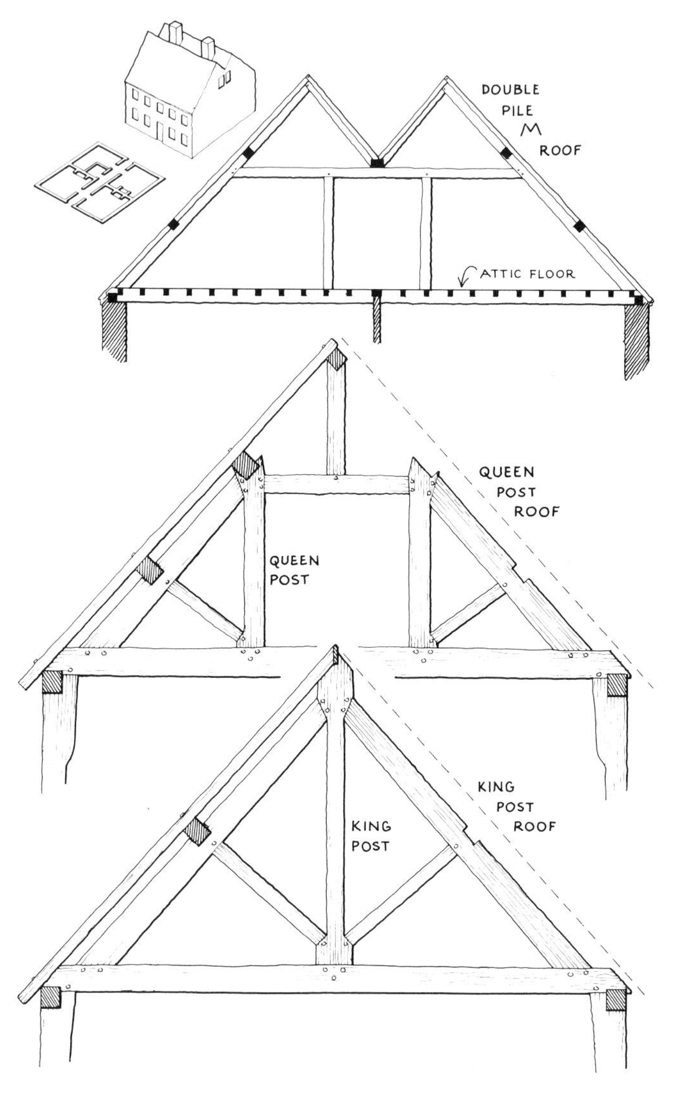

Fig. 44. Eighteenth-century roof trusses.

nineteenth centuries began to make use of pattern books which showed new and recommended designs, particularly of roof trusses. The influence of technical manuals only became apparent very slowly in everyday buildings, but the changes can be seen, particularly in the use of bolts and straps rather than the traditional pegged mortice and tenon joint.

Roof trusses

In brick buildings the roof purlins sometimes rest directly on brick cross walls, but in many buildings framed roof trusses were still used to support the purlins and tie the buildings together. The simplest form, which was in widespread use, consisted simply of a tie beam with angled struts to support the purlins. Similar trusses had been used in earlier buildings with clasped purlins (fig. 33), but the carpentry became progressively simpler, with the struts simply spiked top and bottom rather than jointed. Roofs built on strutted purlins are still part of standard rationalised-traditional building practice. Another technique, commonly found in the north, was to use the simplest possible roof truss, consisting only of a tie beam and heavy principal rafters, the roof often having a low pitch designed for slates.

Many other framed roof trusses of the late eighteenth or nineteenth century are of king-post or queen-post type. We have already met **king-post** trusses in medieval Yorkshire: essentially they consist of a central post – the king post – which rises from the tie beam to support the ridge of the roof (figs. 41 and 44). Structurally, king-post trusses are very efficient, and they are suitable for quite wide spans because the king post itself acts to prevent the tie beam from sagging. This means that the joint between the king post and the tie beam is in tension. Carpenters evidently realised this: the mortice and tenon joint is often either lengthened and given an extra peg or formed in a dovetail shape, so that, once wedged in place, the tenon cannot withdraw (fig. 39). In many nineteenth-century examples a bolt or strap replaces the mortice and tenon joint altogether.

No one has yet explained the gradual adoption of king-post trusses in all parts of the country in the eighteenth and nineteenth centuries. It cannot have been solely due to pattern books, because examples were already appearing in the late seventeenth century, far away from Yorkshire. A similar mystery surrounds **queen-post** trusses. These are known in some medieval buildings, particularly in Suffolk, where they appear as a kind of aisle-derivative structure: vertical posts support the main roof plate or purlin, but instead of reaching to the ground, as aisle posts, they are supported on the tie beam. In eighteenth- and nineteenth-century examples the queen posts take on a shape similar to king posts, supporting the purlins and jointed by a collar (fig. 44). Again, the joint between the tie beam and the queen posts is often enlarged or replaced by a bolt or strap. Many examples of queen-post trusses have interrupted tie beams (fig. 39) and are admirably suited to this application.

The roofing of **double-pile** buildings (buildings one and a half or two rooms deep) also gave rise to problems for carpenters from the late seventeenth century. If the roofing material was slate a low roof pitch could be adopted and the purlins strutted off internal walls or beams, but if tiles were to be used the resulting roof would rise to an excessive height because of the steep pitch required. One solution commonly adopted was to use two identical pitched roofs side by side, but the resulting central valley divided the roof space into two halves. A useful compromise was to form the roof in an M shape, raising the valley above head height. The carpentry involved was fairly simple, a single collar being used to tie the two halves of the roof together (fig. 44). Other details, such as the purlin system, followed the locally established pattern.

As well as these changes in structural forms, a number of interesting details gradually became established during the eighteenth and nineteenth centuries, such as the change from a heavy ridge beam to the use of a plank ridge (shown in the king-post truss in fig. 44), and the change from framed wind braces to the use of diagonally set rafters to brace the roof longitudinally. Another detail which was well established by the late eighteenth century was the use of corner and dragon tie-beam assemblies in the construction of hipped roofs.

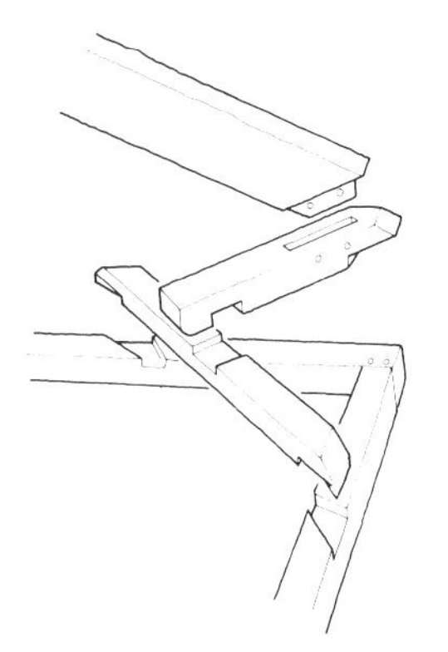

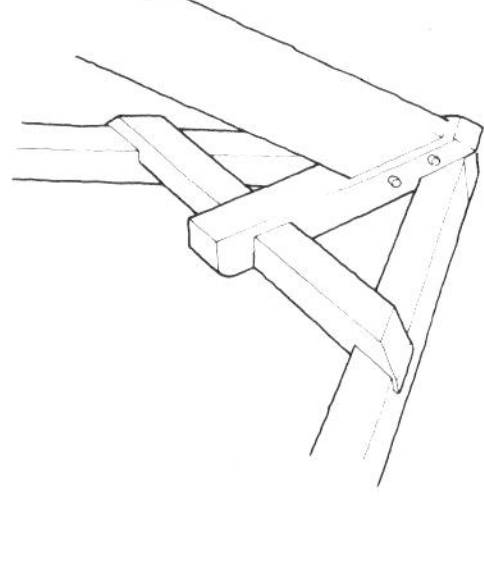

Fig. 45. Hip construction with dragon- and cross-ties.

9. Buildings to visit

There are many timber-framed buildings which are open to the public, and even more which are not timber-framed but contain good traditional carpentry in their roofs, floors or screens. It is not possible here to give anything like a complete list, but the following notes suggest a few examples in which particular features can be easily seen.

Barns are often the best place to examine timber framing. Those which are open to the public tend to be the oldest, largest and most impressive examples, such as those at Great Coxwell (Oxfordshire) and Bredon (Worcestershire), which are aisled barns of the thirteenth century, and at Bradford-on-Avon (Wiltshire), which is of a rather unusual pattern of cruck construction. Other, later, tithe barns illustrate regional construction patterns quite well. Court Lodge, at Brook, Kent, which houses the Wye College Agricultural Museum, is an excellent Kentish aisled barn, and the tithe barn at Harmondsworth (Greater London) is a fifteenth-century aisled structure with a side purlin roof. In West Yorkshire the tithe barn at East Riddlesden Hall, near Keighley, is a good example of a Yorkshire aisled barn with a king-post roof, probably of the sixteenth century. There are at least two smaller barns open to the public. The barn from Hambrook, now in the Weald and Downland Open Air Museum at Singleton, West Sussex, is a small eighteenth-century aisled barn of a type common in West Sussex and east Hampshire; and at Avoncroft Museum of Buildings, Worcestershire, there is a three-bay cruck barn from Cholstrey, Herefordshire, probably sixteenth-century in date.

The other building-type in which timber construction can be seen particularly well is the medieval open hall, and several of these are open to the public in something like their original form. The earliest and grandest examples of timber roofs tend to be in buildings with stone outside walls, such as Stokesay Castle (Salop), Penshurst Place (Kent) and Bowhill (Exeter). At Avoncroft Museum of Buildings an early fourteenth-century roof from the Guesten Hall of Worcester Cathedral can be examined at close quarters as it has been repaired and put back together at ground level, for the time being.

One of the most interesting fully timber-framed medieval halls is West Bromwich Manor (West Midlands), a base cruck structure of about 1300, with which are also associated later wings and a gatehouse, all timber-framed and of high quality. Other fourteenth-century halls which may be mentioned include the White House, Aston Munslow (Salop), and Southchurch Hall, Southend-on-Sea (Essex). In the White House the hall structure is centred on a jointed-cruck truss (rather more grand but on the same principle as the example in fig. 40). Southchurch Hall has a crown-post roof with elaborate cusping under the tie beam and braces of the central truss.

Quite a number of fifteenth-century open halls can be seen in a well-preserved state. At Avoncroft Museum of Buildings the Bromsgrove House is a fairly small merchant's house of the late fifteenth century: the timbers are plain rather than ornamented, and there is an upper floor over the lower bay of the hall, originally perhaps giving a chamber for storage (not a minstrel's gallery). The adjoining two-bay solar cross-wing is probably of the same date as the hall. In the sixteenth century a timber-framed chimney was inserted into the hall, and has been retained in the reconstruction of the building. The original service bay at the lower end of the hall no longer exists. At Singleton Open Air Museum there are two medieval halls. The smaller of the two, Winkhurst, consists simply of a two-bay hall with a chamber above one of the bays, the other being open to the roof. It has a crown-post roof. The other hall, Bayleaf, is an excellent Wealden house (see fig. 34) in which the accommodation can be seen exactly as first designed.

In West Yorkshire, Shibden Hall, Halifax, is a good fifteenth-century example which retains its open hall, as is Lees Hall, Thornhill, near Dewsbury, in which there still exists the canopy over the dais. In Lancashire, Rufford Old Hall, *c.* 1480, is a fine timber hall with a hammer-beam roof, a spere truss and a rare moveable screen. Cruck halls are rarely accessible, unfortunately. Two examples in the north show the crucks to advantage, but the plan arrangement is more obscure: Newton Hall, Hyde, Greater Manchester, and Harome Hall in the Ryedale Folk Museum, Hutton-le-Hole (North Yorkshire).

To discover medieval town buildings the first place to visit is Coventry, where Spon Street is being preserved as a medieval street with the restoration of the existing examples there and the reconstruction of medieval timber-framed buildings displaced by redevelopment in other parts of the city, notably Much Park Street. Of the five buildings so far completed only one retains its open hall: 9 Spon Street, moved from 7 Much Park Street (shown in fig. 28). The original open hearth, revealed by excavation after the building was dismantled, has also been moved and is visible in the open hall.

Another excellent example of a small medieval urban hall can be seen in Church Street, Tewkesbury (Gloucestershire), where a terrace of twenty-two identical units has survived almost intact, each consisting of a shop and a chamber, with a tiny hall and workshop behind. One of the units has been restored to its original form and can be visited. Amongst several other examples which could be mentioned, the Red Lion in Southampton is well worth a visit: it retains its open hall with crown-post roof and a side gallery in the hall for access from front to back of the property. At the other end of the country, the Ark Museum in Tadcaster (North Yorkshire) also has an open hall, with a fifteenth-century king-post roof.

Rufford Old Hall has already been mentioned, but the other fifteenth- and sixteenth-century Lancashire and Cheshire halls with ornate framing form a wonderful group to visit: Speke Hall, Hall-i'-th'-wood, Smithills Hall, Bramall Hall, Gawsworth Hall, Little Moreton Hall and Samlesbury. Another Cheshire house, Churches Mansion at Nantwich, slightly less exuberantly decorated and rather later in date (1577) than most of these, can be visited and enjoyed for, amongst many other things, its carpenters' marks – both arabic and roman, and including some exceptionally long series of numbers.

Rich display framing of a different kind is exhibited by countless buildings in Essex and Suffolk, and in important buildings elsewhere: close studding with moulded and carved beams and posts, the main elevations being almost always jettied. A good example of this in another area, little known but easily accessible, is the Gatehouse at Bolton Percy (North Yorkshire), dated by dendrochronology to 1467; and an example which retains its original (although much restored) painted exterior and moulded plaster infill panels is the Old White Hart in the Market Place at Newark (Nottinghamshire).

Guildhalls, moot halls, wool halls, market halls and the like exist in many timber-framed towns. Important early examples are in York, Leicester, Southampton and Winchester. Essex and Suffolk and the other counties of eastern and south-eastern England possess most of the fifteenth- and sixteenth-century examples, and there are important open market halls in Herefordshire. The market hall from Titchfield (Hampshire) and a guildhall from Crawley (West Sussex) have been reconstructed at the Singleton Open Air Museum.

Timber churches are rare and always worth making a detour to see. Most are in the west and north-west midlands, for instance those at Lower Peover and Baddisley (Cheshire), Melverley (Salop) and Besford (Worcestershire). There are also three in Hampshire, of which that at Hartley Wespall is the most impressive, with marvellous cusped framing in the west-end gable and a superb crown-post roof, dating from the early fourteenth century.

10. Drawing and recording

The best way to study and understand timber-framed buildings is to draw them. There are many ways of drawing, but only one is to be avoided: incomprehensible scribble on the back of an envelope. The clarity of a drawing is a direct reflection of the clarity of the observation behind it. Poor observation will produce poor drawings but, equally, trying to improve your drawings is a useful trick for improving your observation! If some feature of a building does not make immediate sense, the mind may gloss over it but the pencil will have to stop and look, whether you are doing a measured drawing or a sketch.

The basic procedure for carrying out a survey of a building is well summarised in the *Illustrated Handbook of Vernacular Architecture* by R. W. Brunskill, but timber-framed buildings are often dauntingly irregular – nothing is square, the walls lean outwards, and all the door and window openings are modern. Where do we begin? There are four separate things to be done. First, try to analyse the plan, section and elevations of the building, making rough notes as you go. Start in the roof and work downwards, finishing with the external elevations. From this analysis decide which parts of the building are historically significant: the timber frame will usually be the earliest part, but later additions and alterations may very well be worth recording also. Second, carefully measure the original frame, concentrating on obtaining a few accurate dimensions rather than a lot of rough ones. The essentials are: the bay lengths; the width of the building and the roof pitch; the heights of the wall frames and of the upper floor; the pattern of wall framing; the sizes of timbers; and details of original features such as doors, windows, staircases and chimneys. If your measurements are taken along original beams, from joint to joint, you will be able to reconstruct on paper the original appearance of the building, however changed its modern appearance may be, because frames were generally made square by the carpenter. Third, record the building as it is, 'warts and all'. If you are lucky enough to have some days to spare, this can be a detailed measured survey. If not, some photographs and sketches (particularly of details) will suffice. Fourth, make a written description of what you have seen and your interpretation of it. This is a most important step but is often forgotten.

If you have devoted hours of painstaking work to a survey, it is a pity to keep it locked away. Send a copy to the National Monuments Record (23 Savile Row, London W1), who will be glad to receive it and will pay you for it, and give a copy to your local county record office, museum, or archaeological society. If you would like to communicate with a wider audience there are several journals which may be able to publish your work – your local museum will be able to advise you about this. In many counties there are study groups, often based on extramural or WEA classes, in which you will be able to meet like-minded enthusiasts and discuss your work with them.

While it is important and interesting to survey and record buildings, it is even better if this work is combined with a parallel investigation of the *documents* which will give you a fuller picture of a building or a village. There are several easily available books which will help you in this, and your local library or record office will be able to guide you.

11. Black and white

The purpose of this book has been to give a short introduction to a fascinating but complicated subject, and the reader who has struggled this far, his head spinning with crown posts, king posts, plates, purlins and rafters, may wonder what use he can make of all this new and unfamiliar information. Is he really any the wiser? 'I won't believe it until I have seen it in black and white' was a cynic's reaction to the news that colour was coming to television, and it could equally apply to this book. The black and white wonders of Little Moreton Hall are there for all to see but (straining the metaphor a little) to understand how it works as well as how it looks will add vivid colour to the picture.

The most pressing and immediate reason for desiring fuller understanding of timber-framed buildings is that a great deal of repair and restoration is carried out with loving care but in complete absence of technical knowledge. Timber joints and frames are really as easy to get right – or wrong – as mixing a load of concrete but they do not come neatly packaged from the builders' merchant, so the concrete usually wins – with disastrous results. This book contains no specific advice on repair techniques except the most important one of all: no repair will be successful unless the building is fully understood technically as well as aesthetically. If you look at it with sufficient care any historic building will itself tell you both what repairs to do and, to a large extent, how to do them.

As with many things, however, the best reason for discovering timber framing is the sheer enjoyment it gives. Finding a rarity is a great but occasional pleasure. Much more often the enjoyment comes from making sense of a building which has lost, through repeated alterations, its original form. Analysing the bay divisions, for example, may reveal the original plan. Finding empty mortices or peg holes may give a clue to a feature which has disappeared – a doorway or window, perhaps, or a part of the building which has been removed. The facades of town buildings are particularly interesting from this point of view – a house which is now a picturesque muddle will almost always turn out to have been originally a carefully designed showpiece. Close and careful observation is all that is needed.

There are countless buildings to be seen and the excitement of discovery never palls, but after a while something else becomes apparent: the geographical and historical patterns which buildings make. Many features have been described in this book in terms of their geographical distribution. In very few cases can this be accounted for in material or physical terms: what it give us is a *cultural* geography, defining areas of cultural grouping in much the same way as dialect or costume but over a greater time-scale and, as buildings are static, with more precision. Similarly, buildings can act as documents for the historian. Individually they may tell a story, whether of King Charles on the run or of one of his subjects on the make, but collectively they make a pattern revealing the economic and social changes which affected whole communities. Prosperous farmhouses imply prosperous farmers; mean hovels imply impoverished labourers.

Finally, though, after the problems of repair, the excitement of archaeology, and the fine points of history and geography, we arrive back at the buildings themselves. Building is a cultural activity as well as a technical one. If success may be measured by the beauty and simplicity of the marriage between the two, the timber-frame tradition achieved it to a remarkable degree.

Further Reading

The literature on traditional buildings is constantly growing. The reading list below gives only the main national studies, but there are many regional and local ones as well. There is an excellent introductory guide in Brunskill's *Illustrated Handbook* (see below), and select bibliographies in Brunskill's other books. The *Bibliography on Vernacular Architecture* by R. de Z. Hall (David & Charles, 1972) and the first of its supplements, the *Current Bibliography of Vernacular Architecture* edited by D. J. H. Michelmore (Vernacular Architecture Group, 1979), provide the best way to explore the literature in more detail.

General

Armstrong, J. R. *Traditional Buildings Accessible to the Public.* 1979.
Barley, M. W. *The English Farmhouse and Cottage.* 1972.
Brunskill, R. W. *Illustrated Handbook of Vernacular Architecture.* 1971.
Brunskill, R. W. *Traditional Buildings of Britain.* 1981.
Brunskill, R. W. *Traditional Farm Buildings of Britain.* 1982.
Clifton Taylor, A. *The Pattern of English Building.* 1972.
Goodman, W. L. *The History of Woodworking Tools.* 1964.
Mercer, E. *English Vernacular Houses.* 1975.
Rose, W. *The Village Carpenter.* 1952.
Salzman, L. F. *Building in England down to 1540.* 1952.
Smith, P. *Houses of the Welsh Countryside.* 1975.

Timber framing

Alcock, N. W. *Cruck Construction, an introduction and catalogue.* 1981.
Brunskill, R. W. *Timber Building in England.* 1985.
Charles, F. W. B. *Conservation of Timber Buildings.* 1984.
Crossley, F. H. *Timber Building in England.* 1951.
Hewett, C. A. *English Historic Carpentry.* 1980.
Hewett, C. A. *English Cathedral and Monastic Carpentry.* 1985.
Innocent, C. F. *The Development of English Building Construction.* 1916 (reprint 1971).
Smith, J. T. 'Timber-framed Building in England,' *Archaeological Journal*, vol. 122, 1966.

Glossary — index

Definitions are given of words not fully explained in the text. Illustrated references are given in **bold**.

Jointed cruck **77,** 88
Joists 18, **26,** 56
Jowl
Also known as 'rootstock': the enlarged head of a main post which permits the tie beam, wall plate and post to be jointed together. Also applied to the enlarged head of any post.
King post **79, 85,** 87
N.B. This term was at one time used to describe what are now known as crown posts, but this usage is now obsolete.
Lap dovetail **13,** 75
Lap joints **13**
Large framing **23**
Wall framing with large rectangular panels.
Ledge **26**
Also known as a clamp: a timber pegged on to the inside face of a timber frame to support floor joists.
Lobby entrance **51**
Lowland zone 11, 61
Mortice and tenon joints **13, 85**
Mouldings **26, 27,** 31, **56, 63,** 89
Decorative profiles cut into the surface of a timber.
Mullion **25**
National Monuments Record 90
Outshot **29**
Pargeting 20
External plaster applied to a timber frame.
Party walls 53
Passing brace, passing shore **68**
A brace which passes across the other vertical or horizontal members in a frame, usually halved or lapped across and pegged to them. 'Passing shore' is the member which rises from the aisle sill beam to the arcade post in many aisled barns.
Plank and stud 77
Plate
A horizontal timber at the head or foot of a frame.
Posts **7**
In wall frames, vertical members which rise the full height of the frame, being either main posts at the bay divisions or intermediate posts within the bay; shorter uprights are known as studs. In roof trusses, king posts, queen posts and crown posts carry longitudinal beams, all other uprights being known as struts.
Post and truss **7, 73**
Principal rafters **7, 8, 65,** 75
Rafters jointed into the ends of a tie beam.
Projecting windows **25,** 56
Purlins **7, 8**
Longitudinal roof timbers, intermediate between wall plate and ridge, carried by roof trusses and giving support to the common rafters.
Queen post roof **85**
Rafters, common rafters **7, 17**
Timbers which run up the roof from wall plate to apex, supported on the wall plate and purlins.
Rail
Short horizontal timber in a timber-framed wall, other than a girding beam. The term 'rail' and 'cross rail' are sometimes also used to describe girding beams.
Recessed-bay halls 52, **55, 65**
Ridge 73, **86**
Also known as ridge piece, ridge beam, ridge purlin. A horizontal timber at the roof apex, supporting the top ends of the rafters.
Scantling 17, 20
The dimensions of a timber. Traditionally 'scantlings' were small timbers such as rafters.
Scarf joints 13
Shops **53,** 89
Sill, sill beam **16**
A sill is the lowest horizontal member of a window frame. A

sill beam, also known as sole plate, is the lowest member of a wall frame, into which the posts and studs of the frame are tenoned.

Single rafter roof 63, 81
A roof consisting only of rafters, or rafters and collars; i.e. without purlins, ridge or crown plate.

Sling brace truss **75**

Small framing
An alternative name for square framing.

Smoke bay **27**

Smoke hood **27,** 80

Soffit
The underside of a beam.

Solar **31,** 88

Spandrel
The triangular space between a brace and the post and beam to which it is jointed.

Spere, spere truss 88
Fixed screens projecting in from the outside walls to screen the hall from the entrance doorways, but leaving a wide central opening from the passage into the hall. The 'spere truss' is the cross frame of which the speres form a part.

Square framing **25, 71, 83**

Struts **63, 85**
All vertical or raking uprights in roof trusses except those which support longitudinal members.

Studs
All vertical timbers in a wall or cross frame which are not main or intermediate posts.

Tenoned purlins 7, **65**

Threaded purlins 7
Purlins supported by being threaded through holes cut in the principal rafters.

Tie beam **7, 13,** 18, 75

Timber **17,** 19, 73

Trenched purlins **7, 73, 85**

Trimmed opening 28
An opening formed in a timber frame, the trimmer receiving the member which would otherwise block the opening. Most often applied to openings trimmed in timber floors for staircases.

Upper cruck 58, **75**

Upper face **14,** 49
Of a beam, the face used by the carpenter as the marking face for joints.

Wattle and daub **20,** 53

Wealden houses **65,** 88

Wind brace **64**
Brace running from principal rafter to purlin, to stiffen the roof structure.